Problem-Solving Cases in Microsoft® Access and Excel, Fourth Annual Edition

JOSEPH A. BRADY,

ELLEN F. MONK

THOMSON
COURSE TECHNOLOGY™

Australia • Canada • Mexico • Singapore • Spain • United Kingdom • United States

THOMSON

COURSE TECHNOLOGY

Problem-Solving Cases in Microsoft® Access and Excel, Fourth Annual Edition

Publisher:
Bob Woodbury

Acquisitions Editor:
Maureen Martin

Senior Product Manager:
Tricia Coia

Development Editor:
DeVona Dors

Senior Marketing Manager:
Karen Seitz

Associate Product Manager:
Jennifer Smith

Editorial Assistant:
Allison Murphy

Production Editor:
Danielle Chouhan

Manufacturing Coordinator:
Justin Palmeiro

Cover Designer:
Laura Rickenbach

Compositor:
Gex Publishing Services

Copyeditor:
Harry Johnson

Proofreader:
Christine Clark

Indexer:
Rich Carlson

Printed in the United States of America

1 2 3 4 5 6 7 8 9 Glob 09 08 07 06

For more information, contact
Course Technology, 25 Thomson Place,
Boston, Massachusetts, 02210.

Or find us on the World Wide Web at:
www.course.com

ISBN 1-4188-3706-7

DEDICATION

To our problem-solving students

Preface

For the past 15 years, we have taught MIS courses at the University of Delaware. From the start, we wanted to use good computer-based case studies for the database and decision-support portions of our courses.

We could not find a casebook that met our needs! This surprised us because our requirements, we thought, were not unreasonable. First, we wanted cases that asked students to think about real-world business situations. Second, we wanted cases that provided students with hands-on experience, using the kind of software that they had learned to use in their computer literacy courses—and that they would later use in business. Third, we wanted cases that would strengthen students' ability to analyze a problem, examine alternative solutions, and implement a solution using software. Undeterred by the lack of casebooks, we wrote our own cases, and Thomson Course Technology published them.

This is the fifth casebook we have written for Thomson Course Technology. The cases are all new and the tutorials are updated. New features of this textbook reflect comments and suggestions that we have received from instructors who have used our casebook in their classrooms.

As with our prior casebooks, we include tutorials that prepare students for the cases, which are challenging but doable. Most of the cases are organized in a way that helps the student think about the logic of each case's business problem and then how to use the software to solve the business problem. The cases will fit well in an undergraduate MIS course, an MBA Information Systems course, or a Computer Science course devoted to business-oriented programming.

Book Organization

The book is organized into six parts:

1. Database Cases Using Access
2. Decision Support Cases Using Excel Scenario Manager
3. Decision Support Cases Using the Excel Solver
4. Decision Support Case Using Basic Excel Functionality
5. Integration Case: Using Access and Excel
6. Presentation Skills

Part 1 begins with two tutorials that prepare students for the Access case studies. Parts 2 and 3 each begin with a tutorial that prepares students for the Excel case studies. All four tutorials provide students with hands-on practice in using the software's more advanced features—the kind of support that other books about Access and Excel do not give to the student. Part 4 asks students to use Excel's basic functionality for decision support. Part 5 challenges students to use both Access and Excel to find a solution to solve a business problem. Part 6 is a tutorial that hones students' skills in creating and delivering an oral presentation to business managers. The next section explores each of these parts in more depth.

Part 1: Database Cases

Using Access

This section begins with two tutorials and then presents five case studies.

Tutorial A: Database Design

This tutorial helps the student to understand how to set up tables to create a database, without requiring students to learn formal analysis and design methods, such as data normalization.

Tutorial B: Microsoft Access

The second tutorial teaches students the more advanced features of Access queries and reports—features that students will need to know to complete the cases.

Cases 1–5

Five database cases follow Tutorials A and B. The students' job is to implement each case's database in Access so form, query, switchboard, and report outputs can help management. The first case is an easier "warm-up" case. The next four cases require a more demanding database design and implementation effort.

Part 2: Decision Support Cases

Using the Excel Scenario Manager

This section has one tutorial and two decision support cases requiring the use of Excel Scenario Manager.

Tutorial C: Building a Decision Support System in Excel

This section begins with a tutorial using Excel for decision support and spreadsheet design. Fundamental spreadsheet design concepts are taught. Instruction on the Scenario Manager, which can be used to organize the output of many "what-if" scenarios, is emphasized.

Cases 6–7

These two cases can be done with or without the Scenario Manager (although the Scenario Manager is nicely suited to them). In each case, students must use Excel to model two or more solutions to a problem. Students then use the outputs of the model to identify and document the preferred solution via a memorandum and, if assigned to do so, an oral presentation.

Part 3: Decision Support Cases

Using the Excel Solver

This section has one tutorial and two decision support cases requiring the use of Excel Solver.

Tutorial D: Building a Decision Support System Using Excel Solver

This section begins with a tutorial about using the Solver, which is a decision support tool for solving optimization problems.

Cases 8–9

Once again, in each case, students use Excel to analyze alternatives and identify and document the preferred solution.

Part 4: Decision Support Cases

Using Basic Excel Functionality

Case 10

The cases continue with one case that uses basic Excel functionality, i.e., the case does not require the Scenario Manager or the Solver. Excel is used to test the student's analytical skills in "what if" analyses.

Part 5: Integration Case

Using Excel and Access

Case 11

This case integrates Access and Excel. This case is included because of a trend toward sharing data among multiple software packages to solve problems.

Part 6: Presentation Skills

Tutorial E: Giving an Oral Presentation

Each case includes an optional presentation assignment that gives students practice in making a presentation to management on the results of their analysis of the case. This section gives advice on how to create oral presentations. It also has technical information on charting and pivot tables, techniques that might be useful in case analyses or as support for presentations. This tutorial will help students to organize their recommendations, to present their solutions in both words and graphics, and to answer questions from the audience. For larger classes, instructors may wish to have students work in teams to create and deliver their presentations—which would model the "team" approach used by many corporations.

Individual Case Design

The format of the eleven cases follows this template.

- Each case begins with a *Preview* of what the case is about and an overview of the tasks.
- The next section, *Preparation*, tells students what they need to do or know to complete the case successfully. (Of course, our tutorials prepare students for the cases!)
- The third section, *Background*, provides the business context that frames the case. The background of each case models situations that require the kinds of thinking and analysis that students will need in the business world.
- The Background sections are followed by the *Assignment* sections, which are generally organized in a way that helps students to develop their analyses.
- The last section, *Deliverables*, lists what the student must hand in: printouts, a memorandum, a presentation, and files on disk. The list is similar to the kind of deliverables that a business manager might demand.

Using the Cases

We have successfully used cases like these in our undergraduate MIS courses. We usually begin the semester with Access database instruction. We assign the Access database tutorials and then a case to each student. Then, for Excel DSS instruction, we do the same thing—assign a tutorial and then a case.

Microsoft Office 2003

Another important feature of the Fourth Annual Edition is the inclusion of a FREE 30-day trial of Microsoft Office 2003 in the back of every book. This CD-ROM contains the entire Microsoft Office 2003 suite.

Technical Information

This textbook was quality assurance tested using the Windows XP Professional operating system, Microsoft Access 2003, and Microsoft Excel 2003.

Data Files and Solution Files

We have created "starter" data files for the Excel cases, so students need not spend time typing in the spreadsheet skeleton. Case 11 also requires students to load a data file. All these files can be found on the Thomson Course Technology Web site, which is available to both students and instructors. Go to *www.course.com* and search for this textbook by title, author, or ISBN. You are granted a license to copy the data files to any computer or computer network used by individuals who have purchased this textbook.

Solutions to the material in the text are available to instructors. These can also be found at *www.course.com*. Search for this textbook by title, author, or ISBN. The solutions are password protected.

Instructor's Manual

An Instructor's Manual is available to accompany this text. The Instructor's Manual contains additional tools and information to help instructors successfully use this textbook. Items such as a Sample Syllabus, Teaching Tips, and Grading Guidelines are an example of the material that can be found in the Instructor's Manual. Instructors should go to *www.course.com* and search for this textbook by title, author, or ISBN. The Instructor's Manual is password protected.

Acknowledgements

We would like to give many thanks to the team at Thomson Course Technology, including to our Developmental Editor, DeVona Dors; Senior Product Manager, Tricia Coia; and to our Production Editor, Danielle Chouhan. As always, we acknowledge our students' diligent work.

Contents

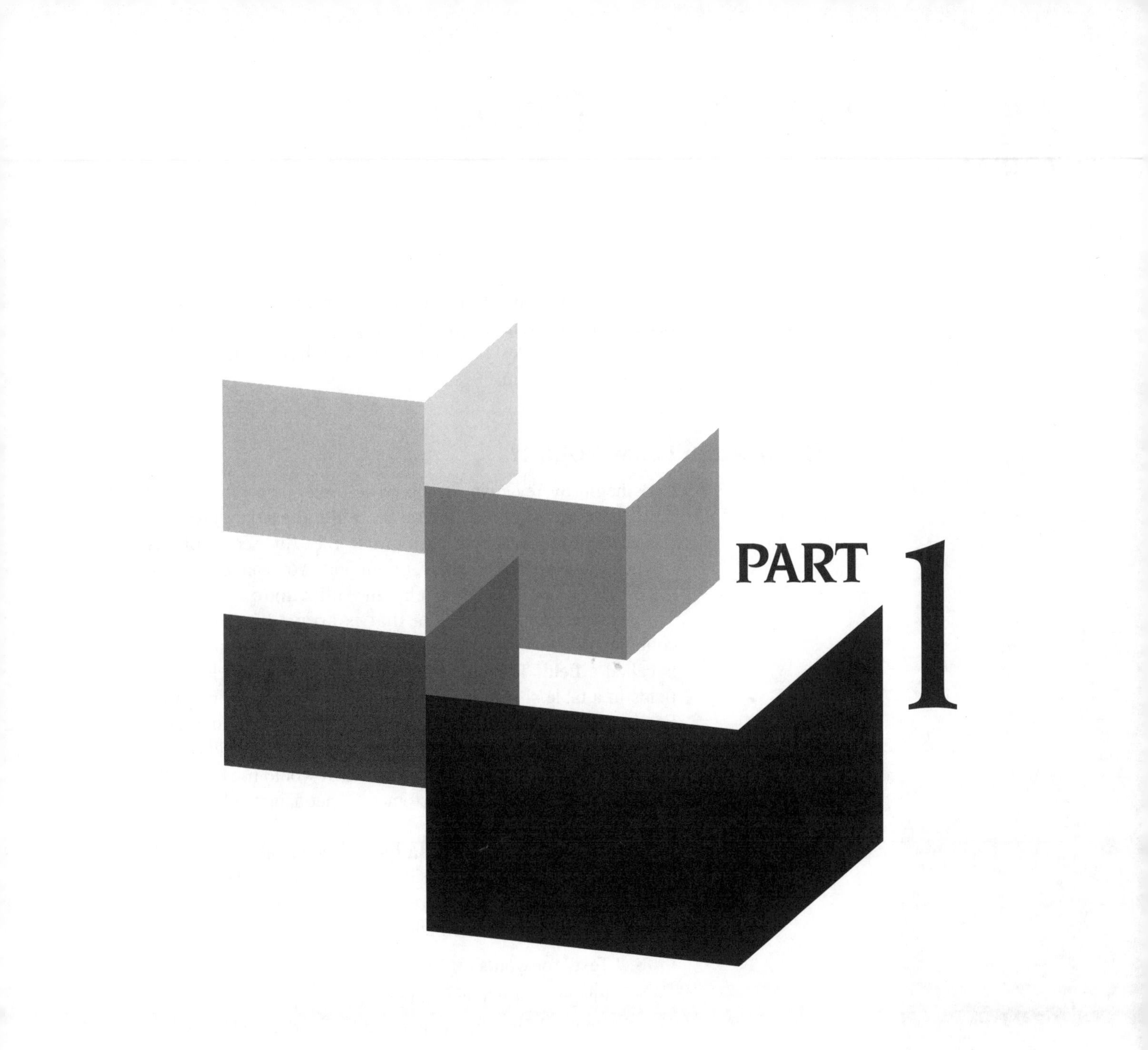

Database Cases Using Access

Database Design

This tutorial has three sections. The first section briefly reviews basic database terminology. The second section teaches database design. The third section has a practice database design problem.

REVIEW OF TERMINOLOGY

Let's begin by reviewing some basic terms that will be used throughout this textbook. In Access, a **database** is a group of related objects that are saved into one file. An Access **object** can be a table, a form, a query, or a report. You can identify an Access database file because it has the suffix **.mdb**.

A **table** consists of data that is arrayed in rows and columns. A **row** of data is called a **record**. A **column** of data is called a **field**. Thus, a record is a set of related fields. The fields in a table should be related to one another in some way. For example, a company might have employee data in a table called EMPLOYEE. That table would contain data fields about employees—their names, addresses, etc. It would not have data fields about the company's customers—that data would go into a CUSTOMER table.

A field's values have a **data type**. When a table is defined, the nature of each field's data is declared. Then, when data is entered, the database software knows how to interpret each entry. Data types in Access include the following:

- "Text" for words
- "Integer" for whole numbers
- "Double" for numbers that can have a decimal value
- "Currency" for numbers that should be treated as dollars and cents
- "Yes/No" for variables that can have only two values (1-0, on/off, yes/no, true/false)
- "Date/Time" for variables that are dates or times

Each database table should have a **primary key** field, a field in which each record has a *unique* value. For example, in an EMPLOYEE table, a field called SSN (for Social Security Number) could be a primary key, because each record's SSN value would be different from every other record's SSN value. Sometimes, a table does not have a single field whose values are all different. In that case, two or more fields are combined into a **compound primary key**. The combination of the fields' values is unique.

Database tables should be logically related to one another. For example, suppose that a company has an EMPLOYEE table with fields for SSN, Name, Address, and Telephone Number. For payroll purposes, the company has an HOURS WORKED table with a field that summarizes Labor Hours for individual employees. The relationship between the EMPLOYEE table and the HOURS WORKED table needs to be established in the database so you can tell which employees worked which hours. This is done by including the primary key field from the EMPLOYEE table (SSN) as a field in the HOURS WORKED table. In the HOURS WORKED table, the SSN field is then called a **foreign key**.

Data can be entered into a table directly or by entering the data into a **form**, which is based on the table. The form then inserts the data into the table.

A **query** is a question that is posed about data in a table (or tables). For example, a manager might want to know the names of employees who have worked for the company more than five years. A query could be designed to interrogate the EMPLOYEE table in that way. The query would be "run" and its output would answer the question.

A query may need to pull data from more than one table, so queries can be designed to interrogate more than one table at a time. In that case, the tables must first be connected by a **join** operation, which links tables on the values in a field that they have in common. The common field acts as a kind of "hinge" for the joined tables; the query generator treats the joined tables as one large table when running the query.

In Access, queries that answer a question are called **select** queries because they select relevant data from the database records. Queries can be designed that will change data in records, add a record to the end of a table, or delete entire records from a table. These are called **update**, **append**, and **delete** queries, respectively.

Access has a **report** generator that can be used to format a table's data or a query's output.

DATABASE DESIGN

"Designing" a database refers to the process of determining which tables need to be in the database and the fields that need to be in each table. This section begins with a discussion of design concepts. The following key concepts are defined:

- Entities
- Relationships
- Attributes

This section then discusses database design rules, a series of steps we advise that you use to build a database.

Database Design Concepts

Computer scientists have formal ways of documenting a database's logic, but learning the notations and mechanics can be quite time-consuming and difficult. Doing this usually takes a good portion of a Systems Analysis and Design course. This tutorial will teach you database design by emphasizing practical business knowledge. This approach will let you design serviceable databases. Your instructor may add some more formal techniques.

A database models the logic of an organization's operation, so your first task is to understand that operation. You do that by talking to managers and workers, by observation, and/or by looking at business documents, such as sales records. Your goal is to identify the business's "entities" (sometimes called *objects*, in yet another use of this term). An **entity** is some thing or some event that the database will contain. Every entity has characteristics, called **attributes**, and a **relationship(s)** to other entities. Let's take a closer look.

Entities

An entity is a tangible thing or an event. The reason for identifying entities is that *an entity eventually becomes a table in the database*. Entities that are things are easy to identify. For example, consider a video store's database. The database would need to contain the names of videotapes and the names of customers who rent them, so you would have one entity named VIDEO and another named CUSTOMER.

By contrast, entities that are events can be more difficult to identify. This is probably because events cannot be seen, but they are no less real. In the video store example, one event would be the VIDEO RENTAL, and another would be HOURS WORKED by employees.

Your analysis is made easier by the knowledge that organizations usually have certain physical entities, such as:

- Employees
- Customers
- Inventory (products or services)
- Suppliers

The database for most organizations would have a table for each of those entities. Your analysis is also made easier by the knowledge that organizations engage in transactions internally and with the outside world. These transactions are the subject of any accounting course, but most people can understand them from events in daily life. Consider the following examples:

- Organizations generate revenue from sales or interest earned. Revenue-generating transactions are event entities, called SALES, INTEREST, etc.
- Organizations incur expenses from paying hourly employees and purchasing materials from suppliers. HOURS WORKED and PURCHASES would be event entities in the databases of most organizations.

Thus, identifying entities is a matter of observing what happens in an organization. Your powers of observation are aided by knowing what entities exist in the databases of most organizations.

Relationships

The analyst should consider the relationship of each entity to other entities. For each entity, the analyst should ask, "What is the relationship, if any, of this entity to every other entity identified?" Relationships can be expressed in English. For example, a college's database might have entities for STUDENT (containing data about each student), COURSE (containing data about each course), and SECTION (containing data about each section). A relationship between STUDENT and SECTION would be expressed as "Students enroll in Sections."

An analyst must also consider what is called the **cardinality** of any relationship. Cardinality can be one-to-one, one-to-many, or many-to-many. These are summarized as follows:

- In a one-to-one relationship, one instance of the first entity is related to just one instance of the second entity.
- In a one-to-many relationship, one instance of the first entity is related to many instances of the second entity, but each instance of the second entity is related to only one instance of the first.

- In a many-to-many relationship, one instance of the first entity is related to many instances of the second entity, and one instance of the second entity is related to many of the first.

To make this more concrete, again think about the college database having STUDENT, COURSE, and SECTION entities. A course, such as Accounting 101, can have more than one section: 01, 02, 03, 04, etc. Thus:

- The relationship between the entities COURSE and SECTION is one-to-many. Each course has many sections, but each section is for just one course.
- The relationship between STUDENT and SECTION is many-to-many. Each student can be in more than one section because each student can take more than one course. Also, each section has more than one student.

Thinking about relationships and their cardinalities may seem tedious to you now. However, you will see that this knowledge will help you to determine the database tables needed (in the case of many-to-many relationships) and the fields that need to be shared between tables (in the case of one-to-many relationships).

Attributes

An attribute is a characteristic of an entity. You identify attributes of an entity because *attributes become a table's fields*. If an entity can be thought of as a noun, an attribute can be thought of as an adjective describing the noun. Continuing with the college database example, again think about the STUDENT entity. Students have names. Thus, Last Name would be an attribute, a field, of the STUDENT entity. First Name would be an attribute as well. The STUDENT entity would have an Address attribute, another field, and so on.

Sometimes, it is difficult to tell the difference between an attribute and an entity. One good way to differentiate them is to ask whether there can be more than one of the possible attribute for each entity. If more than one instance is possible, and you do not know in advance how many there will be, then it's an entity. For example, assume that a student could have two (but no more) Addresses—one for "home" and one for "on campus." You could specify attributes Address 1 and Address 2. On the other hand, what if the number of student addresses could not be stipulated in advance, but all addresses had to be recorded? You would not know how many fields to set aside in the STUDENT table for addresses. You would need a STUDENT ADDRESSES table, which could show any number of addresses for a student.

DATABASE DESIGN RULES

Your first task in database design is always to understand the logic of the business situation. You then build a database for the requirements of that situation. To create a context for learning about database design, let's first look at a hypothetical business operation and its database needs.

Example: The Talent Agency

Suppose that you have been asked to build a database for a talent agency. The agency books bands into nightclubs. The agent needs a database to keep track of the agency's transactions and to answer day-to-day questions. Many questions arise in running the business. For example, a club manager might want to know which bands are available on a certain date at a certain time or the agent's fee for a certain band. Similarly, the agent might want to see a list of all band members and the instrument each plays, or a list of all the bands having three members.

Suppose that you have talked to the agent and have observed the agency's business operation. You conclude that your database would need to reflect the following facts:

1. A "booking" is an event in which a certain band plays in a particular club on a particular date, starting at a certain time, ending at a certain time, and for a specific fee. A band can play more than once a day. The Heartbreakers, for example, could play at the East End Cafe in the afternoon and then at the West End Cafe that night. For each booking, the club pays the talent agent, who keeps a 5% fee and then gives the rest to the band.
2. Each band has at least two members and an unlimited maximum number of members. The agent notes a telephone number of just one band member, which is used as the band's contact number. No two bands have the same name or telephone number.
3. No band members in any of the bands have the same name. For example, if there is a Sally Smith in one band, there is no Sally Smith in any other band.
4. The agent keeps track of just one instrument that each band member plays. "Vocals" is an instrument for this record-keeping purpose.
5. Each band has a desired fee. For example, the Lightmetal band might want $700 per booking and would expect the agent to try to get at least that amount for the band.
6. Each nightclub has a name, an address, and a contact person. That person has a telephone number that the agent uses to contact the club. No two clubs have the same name, contact person name, or telephone number. Each club has a target fee. The contact person will try to get the agent to accept that amount for a band's appearance.
7. Some clubs will feed the band members for free, and others will not.

Before continuing, you might try to design the agency's database on your own. What are the entities? Recall that databases usually have CUSTOMER, EMPLOYEE, and INVENTORY entities and an entity for the revenue-generating transaction event. Each entity becomes a table in the database. What are the relationships between entities? For each entity, what are its attributes? These become the fields in each table. For each table, what is the primary key?

Six Database Design Rules

Assume that you have gathered information about the business situation in the talent agency example. Now you want to identify the tables for the database and then the fields in each table. To do that, observe the following six rules.

Rule 1: You do not need a table for the business itself. The database represents the entire business. Thus, in our example, Agent and Agency are not entities.

Rule 2: Identify the entities in the business description. Look for the things and events that the database must contain. These become tables in the database. Typically, certain entities are represented. In the talent agency example, you should be able to see these entities:

- *Things*: The product (inventory for sale) is Band. The customer is Club.
- *Events*: The revenue-generating transaction is Bookings.

You might ask yourself: Is there an EMPLOYEE entity? Also, isn't INSTRUMENT an entity? These issues will be discussed as the rules are explained.

Rule 3: Look for relationships between the entities. Look for one-to-many relationships between entities. The relationship between these entities must be established in tables, and this is done by using a foreign key. The mechanics of that are discussed in the next rule. (See the discussion of the relationship between Band and Band Member.)

Look for many-to-many relationships between entities. In each of these relationships, there is the need for a third entity that associates the two entities in the relationship. Recall the STUDENT—SECTION many-to-many relationship example. A third table is needed to show the ENROLLMENT of specific students in specific sections. The mechanics of doing this are discussed in the next rule. (See the discussion of the relationship between BAND and CLUB.) (Note that ENROLLMENT can also be thought of as an event entity, and you might have already identified this entity. Forcing yourself to think about many-to-many relationships means that you will not miss it.)

Rule 4: Look for attributes of each entity, and designate a primary key. Think of entities as nouns. List the adjectives of the nouns. These are the attributes which, as was previously mentioned, become the table's fields. After you have identified fields for each table, designate one as the primary key field, if one field has unique values. Designate a compound primary key if no one field has unique values.

The attributes, or fields, of the BAND entity are Band Name, Band Phone Number, and Desired Fee. No two band names can be the same, it is assumed, so the primary key field in this case can be Band Name. Figure A-1 shows the BAND table and its fields: Band Name, Band Phone Number, and Desired Fee; the data type of each field is also shown.

BAND	
Field	***Data Type***
Band Name (primary key)	Text
Band Phone Number	Text
Desired Fee	Currency

Figure A-1 The BAND table and its fields

Two BAND records are shown in Figure A-2.

Band Name (primary key)	*Band Phone Number*	*Desired Fee*
Heartbreakers	981 831 1765	$800
Lightmetal	981 831 2000	$700

Figure A-2 Records in the BAND table

If there could be two bands called the Heartbreakers, then Band Name would not be a good primary key. Some other unique identifier would be needed. Such situations are common. Most businesses have many types of inventory, and duplicate names are possible. The typical solution is to assign a number to each product to be used as the primary key field. For example, a college could have more than one faculty member with the same name, so each faculty member would be assigned an Employee Identification Number (EIN). Similarly, banks assign a Personal Identification Number (PIN) for each depositor. Each automobile that a car manufacturer makes gets a unique Vehicle Identification Number (VIN). Most businesses assign a number to each sale, called an invoice number. (The next time you buy something at a grocery store, note the number on your receipt. It will be different from the number that the next person in line sees on their receipt.)

At this point, you might ask why Band Member would not be an attribute of BAND. The answer is that you must record each band member, but you do not know in advance how many members will be in each band. Therefore, you do not know how many fields to allocate to the BAND table for members. Another way to think about Band Member(s) is that they are, in effect, the agency's employees. Databases for organizations usually have an EMPLOYEE entity. Therefore, you should create a BAND MEMBER table with the attributes Member Name, Band Name, Instrument, and Phone. The BAND MEMBER table and its fields are shown in Figure A-3.

BAND MEMBER	
Field Name	***Data Type***
Member Name (primary key)	Text
Band Name (foreign key)	Text
Instrument	Text
Phone	Text

Figure A-3 The BAND MEMBER table and its fields

Five records in the BAND MEMBER table are shown in Figure A-4.

Member Name (primary key)	***Band Name***	***Instrument***	***Phone***
Pete Goff	Heartbreakers	Guitar	981 444 1111
Joe Goff	Heartbreakers	Vocals	981 444 1234
Sue Smith	Heartbreakers	Keyboard	981 555 1199
Joe Jackson	Lightmetal	Sax	981 888 1654
Sue Hoopes	Lightmetal	Piano	981 888 1765

Figure A-4 Records in the BAND MEMBER table

Instrument can be included as a field in the BAND MEMBER table, because the agent records only one for each band member. Instrument can thus be thought of as a way to describe a band member, much as the phone number is part of the description. Member Name can be the primary key because of the (somewhat arbitrary) assumption that no two members in any band have the same name. Alternatively, Phone could be the primary key if it could be assumed that no two members share a telephone. Or, a band member ID number could be assigned to each person in each band, which would create a unique identifier for each band member handled by the agency.

You might ask why Band Name is included in the BAND MEMBER table. The commonsense reason is that you did not include the Member Name in the BAND table. You must relate bands and members somewhere, and this is the place to do it.

Another way to think about this involves the cardinality of the relationship between BAND and BAND MEMBER. It is a one-to-many relationship: One band has many members, but each member is in just one band. You establish this kind of relationship in the database by using the primary key field of one table as a foreign key in the other. In BAND MEMBER, the foreign key Band Name is used to establish the relationship between the member and his or her band.

The attributes of the entity CLUB are Club Name, Address, Contact Name, Club Phone Number, Preferred Fee, and Feed Band? The table called CLUB can define the CLUB entity, as shown in Figure A-5.

CLUB	
Field Name	***Data Type***
Club Name (primary key)	Text
Address	Text
Contact Name	Text
Club Phone Number	Text
Preferred Fee	Currency
Feed Band?	Yes/No

Figure A-5 The CLUB table and its fields

Two records in the CLUB table are shown in Figure A-6.

Club Name (primary key)	*Address*	*Contact Name*	*Club Phone Number*	*Preferred Fee*	*Feed Band?*
East End	1 Duce St.	Al Pots	981 444 8877	$600	Yes
West End	99 Duce St.	Val Dots	981 555 0011	$650	No

Figure A-6 Records in the CLUB table

You might wonder why Bands Booked Into Club (or some such field name) is not an attribute of the CLUB table. There are two answers. First, you do not know in advance how many bookings a club will have, so the value cannot be an attribute. Furthermore, BOOKINGS is the agency's revenue-generating transaction, an event entity, and you need a table for that business transaction. Let us consider the booking transaction next.

You know that the talent agent books a certain band into a certain club on a certain date, for a certain fee, starting at a certain time, and ending at a certain time. From that information, you can see that the attributes of the BOOKINGS entity are Band Name, Club Name, Date, Start Time, End Time, and Fee. The BOOKINGS table and its fields are shown in Figure A-7.

BOOKINGS	
Field Name	***Data Type***
Band Name	Text
Club Name	Text
Date	Date/Time
Start Time	Date/Time
End Time	Date/Time
Fee	Currency

Figure A-7 The BOOKINGS table and its fields—and no designation of a primary key

Some records in the BOOKINGS table are shown in Figure A-8.

Band Name	*Club Name*	*Date*	*Start Time*	*End Time*	*Fee*
Heartbreakers	East End	11/21/05	19:00	23:30	$800
Heartbreakers	East End	11/22/05	19:00	23:30	$750
Heartbreakers	West End	11/28/05	13:00	18:00	$500
Lightmetal	East End	11/21/05	13:00	18:00	$700
Lightmetal	West End	11/22/05	13:00	18:00	$750

Figure A-8 Records in the BOOKINGS table

No single field is guaranteed to have unique values, because each band would be booked many times, and each club would be used many times. Further, each date and time could appear more than once. Thus, no one field can be the primary key.

If a table does not have a single primary key field, you can make a compound primary key whose field values together will be unique. Because one band can be in only one place at a time, one possible solution is to create a compound key consisting of the fields Band Name, Date, and Start Time. An alternative solution is to create a compound primary key consisting of the fields Club Name, Date, and Start Time.

A way to avoid having a compound key would be to create a field called Booking Number. Each booking would get its own unique number, similar to an invoice number.

Here is another way to think about this event entity: Over time, a band plays in many clubs, and each club hires many bands. The BAND-to-CLUB relationship is, thus, a many-to-many relationship. Such relationships signal the need for a table between the two entities in the relationship. Here, you need the BOOKINGS table that associates the BAND and CLUB tables. An associative table is implemented by including the primary keys from the two tables that are associated. In this case, the primary keys from the BAND and CLUB tables are included as foreign keys in the BOOKINGS table.

Rule 5: Avoid data redundancy. You should not include extra (redundant) fields in a table. Doing this takes up extra disk space, and it leads to data entry errors because the same value must be entered in multiple tables, and the chance of a keystroke error increases. In large databases, keeping track of multiple instances of the same data is nearly impossible, and contradictory data entries become a problem.

Consider this example: Why wouldn't Club Phone Number be in the BOOKINGS table as a field? After all, the agent might have to call about some last-minute change for a booking and could quickly look up the number in the BOOKINGS table. Assume that the BOOKINGS table had Booking Number as the primary key and Club Phone Number as a field. Figure A-9 shows the BOOKINGS table with the unnecessary field.

BOOKINGS	
Field Name	***Data Type***
Booking Number (primary key)	Text
Band Name	Text
Club Name	Text
Club Phone Number	Text
Date	Date/Time
Start Time	Date/Time
End Time	Date/Time
Fee	Currency

Figure A-9 The BOOKINGS table with an unnecessary field—Club Phone Number

The fields Date, Start Time, End Time, and Fee logically depend on the Booking Number primary key—they help define the booking. Band Name and Club Name are foreign keys and are needed to establish the relationship between the tables BAND, CLUB, and BOOKINGS. But what about Club Phone Number? It is not defined by the Booking Number. It is defined by Club Name—*that is, it's a function of the club, not of the booking.* Thus, the Club Phone Number field does not belong in the BOOKINGS table. It's already in the CLUB table, and if the agent needs it, he can look it up there.

Perhaps you can see the practical data entry problem with including Club Phone Number in BOOKINGS. Suppose that a club changed its contact phone number. The agent can easily change the number one time, in CLUB. But now the agent would need to remember the names of all the other tables that have that field as well, and change the values there too. Of course, with a small database, that might not be a difficult thing to recall. But in large databases having many redundant fields in many tables, this sort of maintenance becomes very difficult, which means that redundant data is often incorrect.

You might object, saying, "What about all those foreign keys? Aren't they redundant?" In a sense, they are. But they are needed to establish the relationship between one entity and another, as discussed previously.

Rule 6: Do not include a field if it can be calculated from other fields. A **calculated field** is made using the query generator. Thus, the agent's fee is not included in the BOOKINGS table because it can be calculated by query (here, 5% times the booking fee).

PRACTICE DATABASE DESIGN PROBLEM

Imagine this scenario: Your town has a library. The library wants to keep track of its business in a database, and you have been called in to build it. You talk to the town librarian, review the old paper-based records, and watch people use the library for a few days. You learn these things about the library:

1. Anyone who lives in the town can get a library card if they ask for one. The library considers each person who gets a card a "member" of the library.
2. The librarian wants to be able to contact members by telephone and by mail. She calls members if their books are overdue or when requested materials become available. She likes to mail a "thank you" note to each member on the yearly anniversary of their joining. Without a database, contacting members can be difficult to do efficiently; for example, there could be more than one member by the name of Sally Smith. Often, a parent and a child have the same first and last name, live at the same address, and share a phone.
3. The librarian tries to keep track of each member's reading "interests." When new books come in, the librarian alerts members whose interests match those books. For example, long-time member Sue Doaks is interested in Western novels, growing orchids, and baking bread. There must be some way to match such a reader's interests with available books. However, although the librarian wants to track all of a member's reading interests, she wants to classify each book as being in just one category of interest. For example, the classic gardening book *Orchids of France* would be classified as a book about orchids or a book about France, but not both.
4. The library stocks many books. Each book has a title and any number of authors. Conceivably, there could be more than one book in the library titled *History of the United States*. Similarly, there could be more than one author with the same name.
5. A writer could be the author of more than one book.
6. A book could be checked out repeatedly as time goes on. For example, *Orchids of France* could be checked out by one member in March, by another in July, and by yet another member in September.
7. The library must be able to identify whether a book is checked out.
8. A member can check out any number of books in a visit. Conceivably, a member could visit the library more than once a day to check out books, and some members do just that.
9. All books that are checked out are due back in two weeks, no exceptions. The "late" fee is 50 cents per day late. The librarian would like to have an automated way of generating an overdue book list each day, so she could telephone the miscreants.
10. The library has a number of employees. Each employee has a job title. The librarian is paid a salary, but other employees are paid by the hour. Employees clock in and clock out each day. Assume that all employees work only one shift per day, and all are paid weekly. Pay is deposited directly into employees' checking accounts—no checks are hand-delivered. The database needs to include the librarian and all other employees.

Design the library's database, following the rules set forth in this tutorial. Your instructor will specify the format for your work. Here are a few hints, in the form of questions:

- A book can have more than one author. An author can write more than one book. How would you describe the relationship between books and authors?
- The library lends books for free, of course. If you thought of checking out a book as a sale, for zero revenue, how would you handle the library's revenue-generating event?
- A member can check out any number of books in a check-out. A book can be checked out more than once. How would you describe the relationship between check-outs and books?

Microsoft Access Tutorial

B

TUTORIAL

Microsoft Access is a relational database package that runs on the Microsoft Windows operating system. This tutorial was prepared using Access 2003.

Before using this tutorial, you should know the fundamentals of Microsoft Access and know how to use Windows. This tutorial teaches you some advanced Access skills you'll need to do database case studies. This tutorial concludes with a discussion of common Access problems and how to solve them.

A preliminary caution: Always observe proper file-saving and closing procedures. Use these steps to exit from Access: (1) File—Close, then (2) File—Exit. This gets you back to Windows. Always end your work with these two steps. Never pull out your disk, CD, or other portable storage device and walk away with work remaining on the screen, or you will lose your work.

To begin this tutorial, you will create a new database called **Employee**.

At the Keyboard

Open a new database (in the Task Pane—New—Blank database). (According to Microsoft, the Task Pane is a universal remote control, which saves the user steps.) Call the database **Employee**. If you are saving to a floppy disk, first select the drive (**A:**), and then enter the filename. **EMPLOYEE.mdb** would be a good choice.

Your opening screen should resemble the screen shown in Figure B-1.

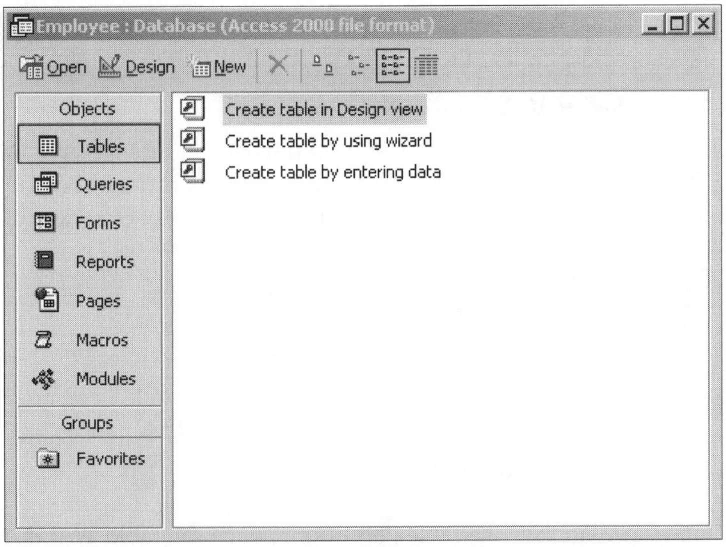

Figure B-1 The Database window in Access

In this tutorial, the screen shown in Figure B-1 is called the Database window. From this screen, you can create or change objects.

Creating Tables

Your database will contain data about employees, their wage rates, and their hours worked.

Defining Tables

In the Database window, make three new tables, using the instructions that follow.

At the keyboard

(1) Define a table called EMPLOYEE.

This table contains permanent data about employees. To create it, in the Table Objects screen, click New, then Design View, and then define the table EMPLOYEE. The table's fields are Last Name, First Name, SSN (Social Security Number), Street Address, City, State, Zip, Date Hired, and US Citizen. The field SSN is the primary key field. Change the length of text fields from the default 50 spaces to more appropriate lengths; for example, the field Last Name might be 30 spaces, and the Zip field might be 10 spaces. Your completed definition should resemble the one shown in Figure B-2.

Table1 : Table

	Field Name	Data Type	Description
	Last Name	Text	
	First Name	Text	
🔑	SSN	Text	
	Street Address	Text	
	City	Text	
	State	Text	
	Zip	Text	
	Date Hired	Date/Time	
	US Citizen	Yes/No	

Figure B-2 Fields in the EMPLOYEE table

When you're finished, choose File—Save. Enter the name desired for the table (here, EMPLOYEE). Make sure that you specify the name of the *table*, not the database itself. (Here, it is a coincidence that the EMPLOYEE table has the same name as its database file.)

(2) Define a table called WAGE DATA.

This table contains permanent data about employees and their wage rates. The table's fields are SSN, Wage Rate, and Salaried. The field SSN is the primary key field. Use the data types shown in Figure B-3. Your definition should resemble the one shown in Figure B-3.

Table1 : Table

	Field Name	Data Type	Description
🔑▶	SSN	Text	
	Wage Rate	Currency	
	Salaried	Yes/No	

Figure B-3 Fields in the WAGE DATA table

Use File—Save to save the table definition. Name the table WAGE DATA.

(3) Define a table called HOURS WORKED.

The purpose of this table is to record the number of hours employees work each week in the year. The table's fields are SSN (text), Week # (number—long integer), and Hours (number—double). The SSN and Week# are the compound keys.

In the following example, the employee having SSN 089-65-9000 worked 40 hours in Week 1 of the year and 52 hours in Week 2.

SSN	Week #	Hours
089-65-9000	1	40
089-65-9000	2	52

Note that no single field can be the primary key field. Why? Notice that 089-65-9000 is an entry for each week. If the employee works each week of the year, at the end of the year, there will be 52 records with that value. Thus, SSN values will not distinguish records. However, no other single field can distinguish these records either, because other employees will have worked during the same week number, and some employees will have worked the same number of hours (40 would be common).

However, a table must have a primary key field. What is the solution? Use a compound primary key; that is, use values from more than one field. Here, the compound key to use consists of the field SSN plus the Week # field. Why? There is only *one* combination of SSN 089-65-9000 and Week# 1—those values *can occur in only one record*; therefore, the combination distinguishes that record from all others.

How do you set a compound key? The first step is to highlight the fields in the key. These must appear one after the other in the table definition screen. (Plan ahead for this format.) Alternately, you can highlight one field, hold down the Control key, and highlight the next field.

AT THE KEYBOARD

For the HOURS WORKED table, click in the first field's left prefix area, hold down the button, then drag down to highlight names of all fields in the compound primary key. Your screen should resemble the one shown in Figure B-4.

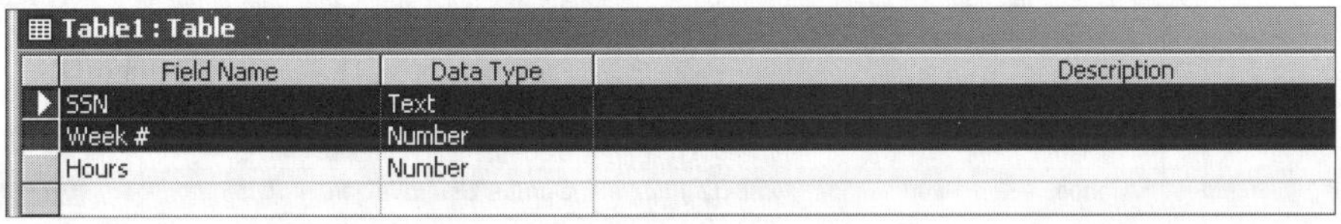

Table1 : Table

Field Name	Data Type	Description
SSN	Text	
Week #	Number	
Hours	Number	

Figure B-4 Selecting fields as the compound primary key for the HOURS WORKED table

Now, click the Key icon. Your screen should resemble the one shown in Figure B-5.

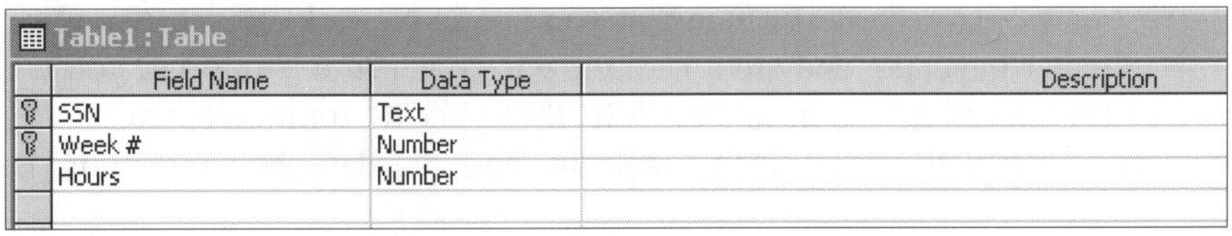

Table1 : Table

Field Name	Data Type	Description
SSN	Text	
Week #	Number	
Hours	Number	

Figure B-5 The compound primary key for the HOURS WORKED table

That completes the compound primary key and the table definition. Use File—Save to save the table as HOURS WORKED.

Adding Records to a Table

At this point, all you have done is to set up the skeletons of three tables. The tables have no data records yet. If you were to print the tables, all you would see would be column headings (the field names). The most direct way to enter data into a table is to select the table, open it, and type the data directly into the cells.

AT THE KEYBOARD

At the Database window, select Tables, then EMPLOYEE. Then select Open. Your data-entry screen should resemble the one shown in Figure B-6.

Employee : Table

Last Name	First Name	SSN	Street Address	City	State	Zip	Date Hired	US Citizen

Figure B-6 The data-entry screen for the EMPLOYEE table

The table has many fields, and some of them may be off the screen, to the right. Scroll to see obscured fields. (Scrolling happens automatically as data is entered.) Figure B-6 has been adjusted to view all fields on one screen.

Type in your data, one field value at a time. Note that the first row is empty when you begin. Each time you finish a value, hit Enter, and the cursor will move to the next cell. After the last cell in a row, the cursor moves to the first cell of the next row, *and* Access automatically saves the record. (Thus, there is no File—Save step after entering data into a table.)

Dates (for example, Date Hired) are entered as "6/15/04" (without the quotation marks). Access automatically expands the entry to the proper format in output.

Yes/No variables are clicked (checked) for Yes; otherwise (for No), the box is left blank. You can click the box from Yes to No, as if you were using a toggle switch.

If you make errors in data entry, click in the cell, backspace over the error, and type the correction.

Enter the data shown in Figure B-7 into the EMPLOYEE table.

Employee : Table

Last Name	First Name	SSN	Street Address	City	State	Zip	Date Hired	US Citizen
Howard	Jane	114-11-2333	28 Sally Dr	Glasgow	DE	19702	8/1/2006	☑
Smith	John	123-45-6789	30 Elm St	Newark	DE	19711	6/1/1996	☑
Smith	Albert	148-90-1234	44 Duce St	Odessa	DE	19722	7/15/1987	☑
Jones	Sue	222-82-1122	18 Spruce St	Newark	DE	19716	7/15/2004	☐
Ruth	Billy	714-60-1927	1 Tater Dr	Baltimore	MD	20111	8/15/1999	☐
Add	Your	Data	Here	Newark	MN	33776		☑

Figure B-7 Data for the EMPLOYEE table

Note that the sixth record is *your* data record. The edit pencil in the left prefix area marks that record. Assume that you live in Newark, Minnesota, were hired on today's date (enter the date), and are a U.S. citizen. (Later in this tutorial, you will see that one entry is for the author's name and the SSN 099-11-3344 for this record.)

Open the WAGE DATA table and enter the data shown in Figure B-8 into the table.

SSN	Wage Rate	Salaried
114-11-2333	$10.00	☐
123-45-6789	$0.00	☑
148-90-1234	$12.00	☐
222-82-1122	$0.00	☑
714-60-1927	$0.00	☑
Your SSN!	$8.00	☐

Figure B-8 Data for the WAGE DATA table

Again, you must enter your SSN. Assume that you earn $8 an hour and are not salaried. (Note that Salaried = No implies someone is paid by the hour. Those who are salaried do not get paid by the hour, so their hourly rate is shown as 0.00.)

Open the HOURS WORKED table and enter the data shown in Figure B-9 into the table.

SSN	Week #	Hours
114-11-2333	1	40
114-11-2333	2	50
123-45-6789	1	40
123-45-6789	2	40
148-90-1234	1	38
148-90-1234	2	40
222-82-1122	1	40
222-82-1122	2	40
714-60-1927	1	40
714-60-1927	2	40
your SSN	1	60
your SSN	2	55

Figure B-9 Data for the HOURS WORKED table

Notice that salaried employees are always given 40 hours. Non-salaried employees (including you) might work any number of hours. For your record, enter your SSN, 60 hours worked for Week 1, and 55 hours worked for Week 2.

Creating Queries

Because you can already create basic queries, this section teaches you the kinds of advanced queries you will create in the Case Studies.

Using Calculated Fields in Queries

A **calculated field** is an output field that is made from *other* field values. A calculated field is *not* a field in a table; it is created in the query generator. The calculated field does not become part of the table—it is just part of query output. The best way to explain this process is by working through an example.

At the Keyboard

Suppose that you want to see the SSNs and wage rates of hourly workers, and you want to see what the wage rates would be if all employees were given a 10% raise. To do this, show the SSN, the current wage rate, and the higher rate (which should be titled New Rate in the output). Figure B-10 shows how to set up the query.

Query1 : Select Query

Wage Data
*
SSN
Wage Rate
Salaried

Field:	SSN	Salaried	Wage Rate	New Rate: 1.1*[Wage Rate]
Table:	Wage Data	Wage Data	Wage Data	
Sort:				
Show:	☑	☐	☑	☑
Criteria:		=No		
or:				

Figure B-10 Query set-up for the calculated field

The Salaried field is needed, with the Criteria =No, to select hourly workers. The Show box for that field is not checked, so the Salaried field values will not show in the query output.

Note the expression for the calculated field, which you see in the rightmost field cell:

New Rate: 1.1*[Wage Rate]

New Rate: merely specifies the desired output heading. (Don't forget the colon.) The 1.1*[Wage Rate] multiplies the old wage rate by 110%, which results in the 10% raise.

In the expression, the field name Wage Rate must be enclosed in square brackets. This is a rule: *Any time that an Access expression refers to a field name, it must be enclosed in square brackets.*

If you run this query, your output should resemble that shown in Figure B-11.

Query1 : Select Query

SSN	Wage Rate	New Rate
114-11-2333	$10.00	11
148-90-1234	$12.00	13.2
099-11-3344	$8.00	8.8

Figure B-11 Output for a query with calculated field

Notice that the calculated field output is not shown in Currency format; it's shown as a Double—a number with digits after the decimal point. To convert the output to Currency format, click the line above the calculated field expression, thus activating the column (it darkens). Your data-entry screen should resemble the one shown in Figure B-12.

Query1 : Select Query

Wage Data: *, SSN, Wage Rate, Salaried

Field:	SSN	Salaried	Wage Rate	New Rate: 1.1*[Wage Rate]
Table:	Wage Data	Wage Data	Wage Data	
Sort:				
Show:	☑	☐	☑	☑
Criteria:		No		
or:				

Figure B-12 Activating a calculated field in query design

Then select View—Properties. Click the Format drop-down menu. A window, such as the one shown in Figure B-13, will pop up.

Field Properties

General | Lookup

Description
Format
Decimal Places
Input Mask
Caption

General Number	3456.78
Currency	$3,456.
Euro	€3,456.
Fixed	3456.79
Standard	3,456.7
Percent	123.00
Scientific	3.46E+

Figure B-13 Field Properties of a calculated field

Click Currency. Then click the upper-right X to close the window. Now when you run the query, the output should resemble that shown in Figure B-14.

Query1 : Select Query

SSN	Wage Rate	New Rate
114-11-2333	$10.00	$11.00
148-90-1234	$12.00	$13.20
099-11-3344	$8.00	$8.80

Figure B-14 Query output with formatted calculated field

Next, let's look at how to avoid errors when making calculated fields.

Avoiding Errors in Making Calculated Fields

Follow these guidelines to avoid making errors in calculated fields:

- Don't put the expression in the *Criteria* cell, as if the field definition were a filter. You are making a field, so put the expression in the *Field* cell.

- Spell, capitalize, and space a field's name *exactly* as you did in the table definition. If the table definition differs from what you type, Access thinks you're defining a new field by that name. Access then prompts you to enter values for the new field, which it calls a "Parameter Query" field. This is easy to debug because of the tag Parameter Query. If Access asks you to enter values for a Parameter, you almost certainly have misspelled a field name in an expression in a calculated field or a criterion.

  Example: Here are some errors you might make for Wage Rate:

  Misspelling: (Wag Rate)

  Case change: (wage Rate / WAGE RATE)

  Spacing change: (WageRate / Wage   Rate)

- Don't use parentheses or curly braces instead of the square brackets. Also, don't put parentheses inside square brackets. You *are* allowed to use parentheses outside the square brackets, in the normal algebraic manner.

  Example: Suppose that you want to multiply Hours times Wage Rate, to get a field called Wages Owed. This is the correct expression:

  Wages Owed: [Wage Rate]*[Hours]

  This would also be correct:

  Wages Owed: ([Wage Rate]*[Hours])

  But it would **not** be correct to leave out the inside brackets, which is a common error:

  Wages Owed: [Wage Rate*Hours]

"Relating" Two (or More) Tables by the Join Operation

Often, the data you need for a query is in more than one table. To complete the query, you must join the tables. One rule of thumb is that joins are made on fields that have common *values,* and those fields can often be key fields. The names of the join fields are irrelevant—the names may be the same, but that is not a requirement for an effective join.

Make a join by first bringing in (Adding) the tables needed. Next, decide which fields you will join. Then, click one field name and hold down the left mouse button while dragging the cursor over to the other field's name in its window. Release the button. Access puts in a line, signifying the join. (*Note*: If there are two fields in the tables with the same name, Access will put in the line automatically, so you do not have to do the click-and-drag operation.)

You can join more than two tables together. The common fields *need not* be the same in all tables; that is, you can "daisy-chain" them together.

A common join error is to Add a table to the query and then fail to link it to another table. You have a table just "floating" in the top part of the QBE screen! When you run the query, your output will show the same records over and over. This error is unmistakable because there is *so much* redundant output. The rules are: (1) add only the tables you need and (2) link all tables.

Next, you'll work through an example of a query needing a join.

At the Keyboard

Suppose that you want to see the last names, SSNs, wage rates, salary status, and citizenship only for U.S. citizens and hourly workers. The data is spread across two tables, EMPLOYEE and WAGE DATA, so both tables are added, and five fields are pulled down. Criteria are then added. Set up your work to resemble that shown in Figure B-15.

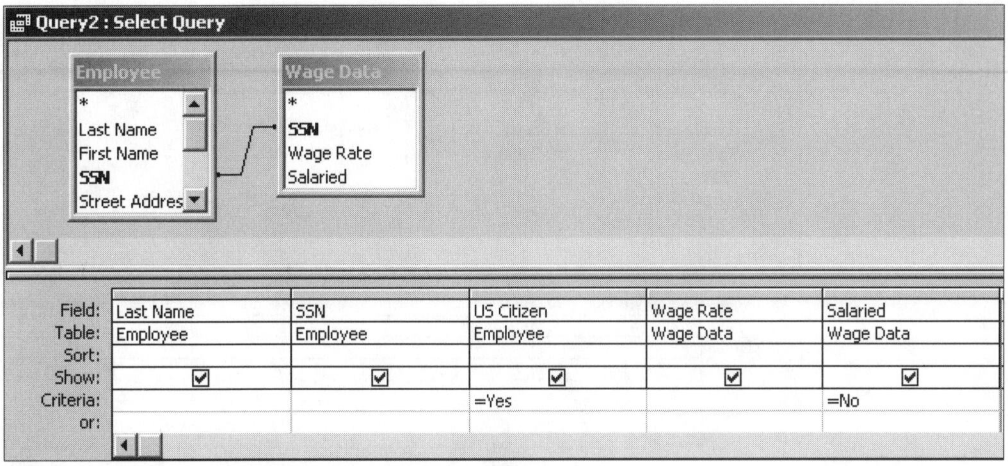

Query2 : Select Query

Field:	Last Name	SSN	US Citizen	Wage Rate	Salaried
Table:	Employee	Employee	Employee	Wage Data	Wage Data
Sort:					
Show:	☑	☑	☑	☑	☑
Criteria:			=Yes		=No
or:					

Figure B-15 A query based on two joined tables

In Figure B-15, the join is on the SSN field. A field by that name is in both tables, so Access automatically puts in the join. If one field had been spelled SSN and the other Social Security Number, you would still join on these fields (because of the common values). You would click and drag to do this operation.

Now run the query. The output should resemble that shown in Figure B-16, with the exception of the name Brady.

Query2 : Select Query

Last Name	SSN	US Citizen	Wage Rate	Salaried
Howard	114-11-2333	☑	$10.00	☐
Smith	148-90-1234	☑	$12.00	☐
Brady	099-11-3344	☑	$8.00	☐

Figure B-16 Output of a query based on two joined tables

Here is a quick review of Criteria: If you want data for employees who are U.S. citizens *and* who are hourly workers, the Criteria expressions go into the *same* Criteria row. If you want data for employees who are U.S. citizens *or* who are hourly workers, one of the expressions goes into the second Criteria row (the one that has the "or:" notation).

There is no need to print the query output or to save it. Go back to the Design View and close the query. Another practice query follows.

AT THE KEYBOARD

Suppose that you want to see the wages owed to hourly employees for Week 2. Show the last name, the SSN, the salaried status, the week #, and the wages owed. Wages will have to be a calculated field ([Wage Rate]*[Hours]). The criteria are =No for Salaried and =2 for the Week # (another "And" query). You'd set up the query the way it is displayed in Figure B-17.

Query2 : Select Query

Field:	Last Name	SSN	Salaried	Week #	Pay: [Wage Rate]*[Hours]
Table:	Employee	Employee	Wage Data	Hours Worked	
Sort:					
Show:	☑	☑	☑	☑	☑
Criteria:			=No	=2	
or:					

Figure B-17 Query set-up for wages owed to hourly employees for Week 2

NOTE

In the previous table, the calculated field column was widened so you can see the whole expression. To widen a column, remember to click the column boundary line and drag to the right.

Run the query. The output should be similar to that shown in Figure B-18 (if you formatted your calculated field to currency).

Query2 : Select Query

Last Name	SSN	Salaried	Week #	Pay
Howard	114-11-2333	☐	2	$500.00
Smith	148-90-1234	☐	2	$480.00
Brady	099-11-3344	☐	2	$440.00

Figure B-18 Query output for wages owed to hourly employees for Week 2

Notice that it was not necessary to pull down the Wage Rate and Hours fields to make this query work. Return to the Design View. There is no need to save. Select File—Close.

Summarizing Data from Multiple Records (Sigma Queries)

You may want data that summarizes values from a field for several records (or possibly all records) in a table. For example, you might want to know the average hours worked for all employees in a week, or perhaps the total (sum of) all the hours worked. Furthermore, you might want data grouped ("stratified") in some way. For example, you might want to know the average hours worked, grouped by all U.S. citizens versus all non-U.S. citizens. Access calls this kind of query a "summary" query, or a **Sigma query**. Unfortunately, this terminology is not intuitive, but the statistical operations that are allowed will be familiar. These operations include the following:

Sum	The total of some field's values
Count	A count of the number of instances in a field, that is, the number of records. Here, to get the number of employees, you'd count the number of SSN numbers.
Average	The average of some field's values

Min	The minimum of some field's values
Var	The variance of some field's values
StDev	The standard deviation of some field's values

AT THE KEYBOARD

Suppose that you want to know how many employees are represented in a database. The first step is to bring the EMPLOYEE table into the QBE screen. Do that now. The query will Count the number of SSNs, which is a Sigma query operation. Thus, you must bring down the SSN field.

To tell Access you want a Sigma query, click the little "Sigma" icon in the menu, as shown in Figure B-19.

Σ

Figure B-19 Sigma icon

This opens up a new row in the lower part of the QBE screen, called the Total row. At this point, the screen would resemble that shown in Figure B-20.

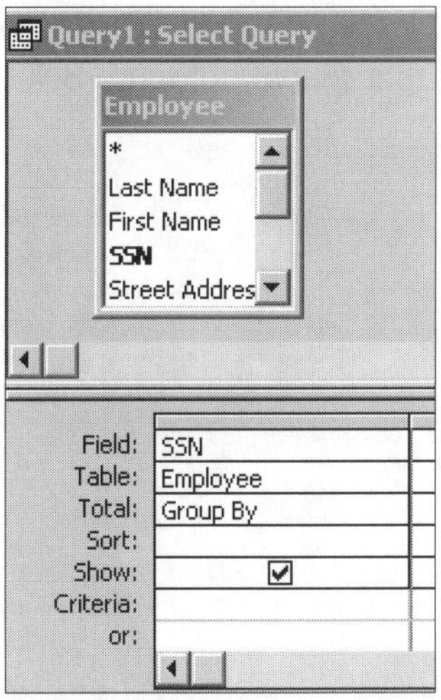

Query1 : Select Query

Employee
*
Last Name
First Name
SSN
Street Addres

Field:	SSN
Table:	Employee
Total:	Group By
Sort:	
Show:	☑
Criteria:	
or:	

Figure B-20 Sigma query set-up

Note that the Total cell contains the words "Group By." Until you specify a statistical operation, Access just assumes that a field will be used for grouping (stratifying) data.

To count the number of SSNs, click next to Group By, revealing a little arrow. Click the arrow to reveal a drop-down menu, as shown in Figure B-21.

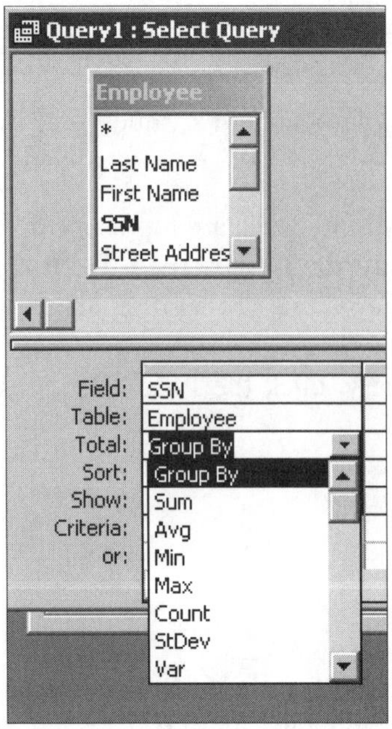

Figure B-21 Choices for statistical operation in a Sigma query

Select the Count operator. (With this menu, you may need to scroll to see the operator you want.) Your screen should now resemble that shown in Figure B-22.

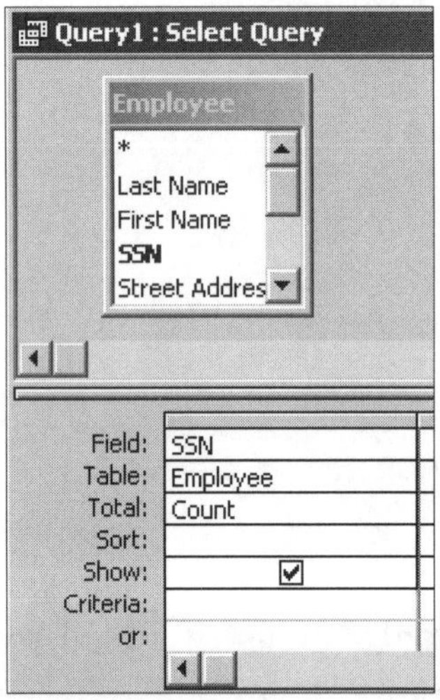

Figure B-22 Count in a Sigma query

Run the query. Your output should resemble that shown in Figure B-23.

Figure B-23 Output of Count in a Sigma query

Notice that Access has made a pseudo-heading "CountOfSSN." To do this, Access just spliced together the statistical operation (Count), the word *Of*, and the name of the field

(SSN). What if you wanted an English phrase, such as "Count of Employees," as a heading? In the Design View, you'd change the query to resemble the one shown in Figure B-24.

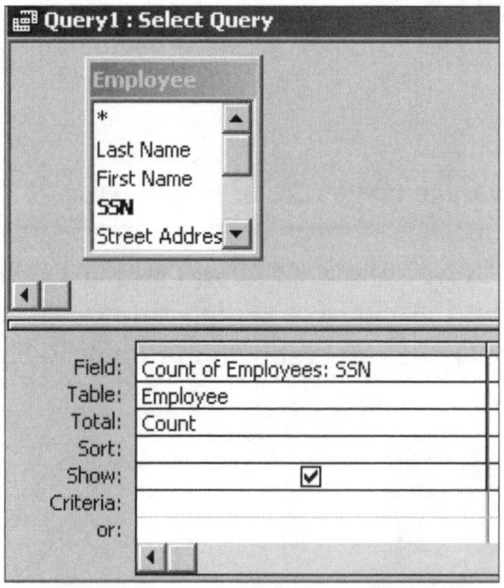

Figure B-24 Heading change in a Sigma query

Now when you run the query, the output should resemble that shown in Figure B-25.

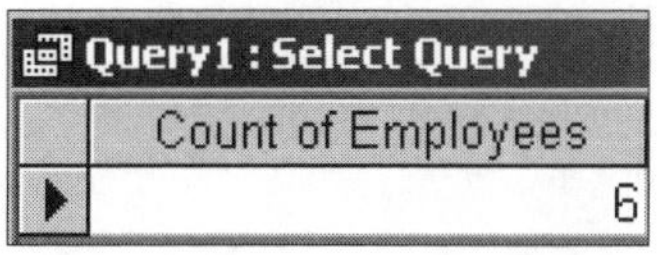

Figure B-25 Output of heading change in a Sigma query

There is no need to save this query. Go back to the Design View and Close.

AT THE KEYBOARD

Here is another example. Suppose that you want to know the average wage rate of employees, grouped by whether they are salaried.

Figure B-26 shows how your query should be set up.

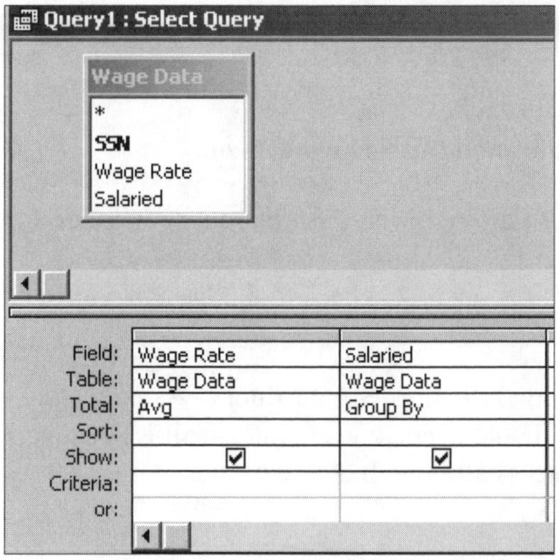

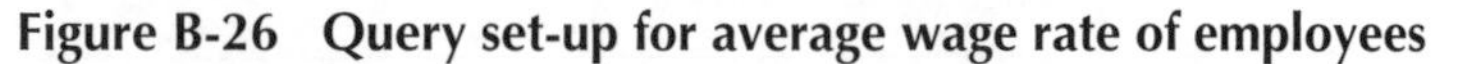

Figure B-26 Query set-up for average wage rate of employees

When you run the query, your output should resemble that shown in Figure B-27.

Query1 : Select Query

AvgOfWage Rate	Salaried
$0.00	☑
$10.00	☐

Figure B-27 Output of query for average wage rate of employees

Recall the convention that salaried workers are assigned zero dollars an hour. Suppose that you want to eliminate the output line for zero dollars an hour because only hourly-rate workers matter for this query. The query set-up is shown in Figure B-28.

Query1 : Select Query

Wage Data
*
SSN
Wage Rate
Salaried

Field:	Wage Rate	Salaried
Table:	Wage Data	Wage Data
Total:	Avg	Group By
Sort:		
Show:	☑	☑
Criteria:		=No
or:		

Figure B-28 Query set-up for non-salaried workers only

When you run the query, you'll get output for non-salaried employees only, as shown in Figure B-29.

Query1 : Select Query

AvgOfWage Rate	Salaried
$10.00	☐

Figure B-29 Query output for non-salaried workers only

Thus, it's possible to use a Criteria in a Sigma query without any problem, just as you would with a "regular" query.

There is no need to save the query. Go back to the Design View and Close.

At the Keyboard

You can make a calculated field in a Sigma query. Assume that you want to see two things for hourly workers: (1) the average wage rate—call it Average Rate in the output; and (2) 110% of this average rate—call it the Increased Rate.

You already know how to do certain things for this query. The revised heading for the average rate will be Average Rate (Average Rate: Wage Rate, in the Field cell). You want the Average of that field. Grouping would be by the Salaried field (with Criteria: =No, for hourly workers).

The most difficult part of this query is to construct the expression for the calculated field. Conceptually, it is as follows:

Increased Rate: 1.1*[The current average, however that is denoted]

The question is how to represent [The current average]. You cannot use Wage Rate for this, because that heading denotes the wages before they are averaged. Surprisingly, it turns out that you can use the new heading (Average Rate) to denote the averaged amount. Thus:

Increased Rate: 1.1*[Average Rate]

Counterintuitively, *you can treat "Average Rate" as if it were an actual field name*. Note, however, that if you use a calculated field, such as Average Rate, in another calculated field, as shown in Figure B-30, you must show that original calculated field in the query output, or the query will ask you to "enter parameter value," which is incorrect. Use the set-up shown in Figure B-30.

Query1 : Select Query

Wage Data
*
SSN
Wage Rate
Salaried

Field:	Average Rate: Wage Rate	Salaried	Increased Rate: 1.1*[Average Rate]
Table:	Wage Data	Wage Data	
Total:	Avg	Group By	Group By
Sort:			
Show:	☑	☑	☑
Criteria:		=No	
or:			

Figure B-30 Using a calculated field in another calculated field

However, if you ran the query now shown in Figure B-30, you'd get some sort of error message. You do not want Group By in the calculated field's Total cell. There is not a *statistical* operator that applies to the calculated field. You must change the Group By operator to Expression. You may have to scroll to get to Expression in the list. Figure B-31 shows how your screen should look.

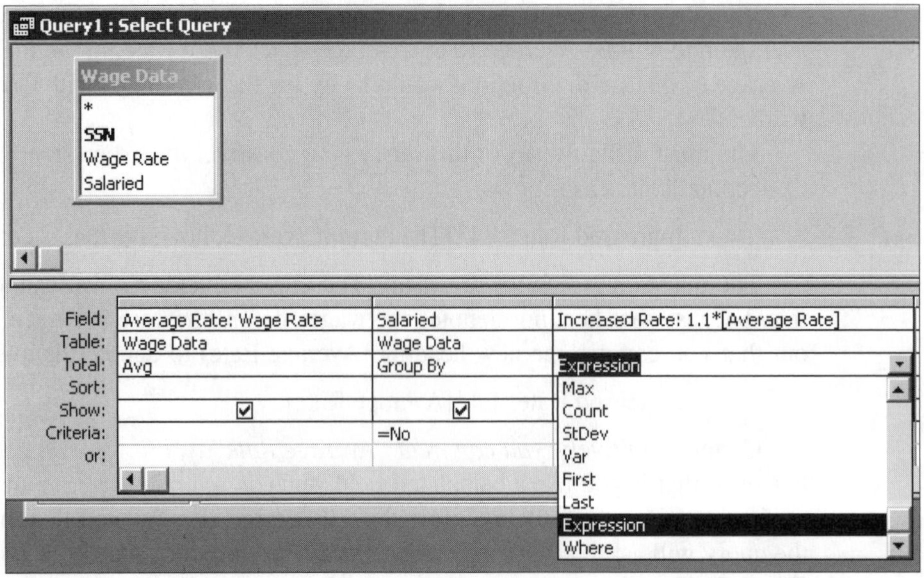

Query1 : Select Query

Wage Data: *, SSN, Wage Rate, Salaried

Field:	Average Rate: Wage Rate	Salaried	Increased Rate: 1.1*[Average Rate]
Table:	Wage Data	Wage Data	
Total:	Avg	Group By	Expression
Sort:			
Show:	☑	☑	
Criteria:		=No	
or:			

Max
Count
StDev
Var
First
Last
Expression
Where

Figure B-31 Changing the Group By to an Expression in a Sigma query

Figure B-32 shows how the screen looks before running the query.

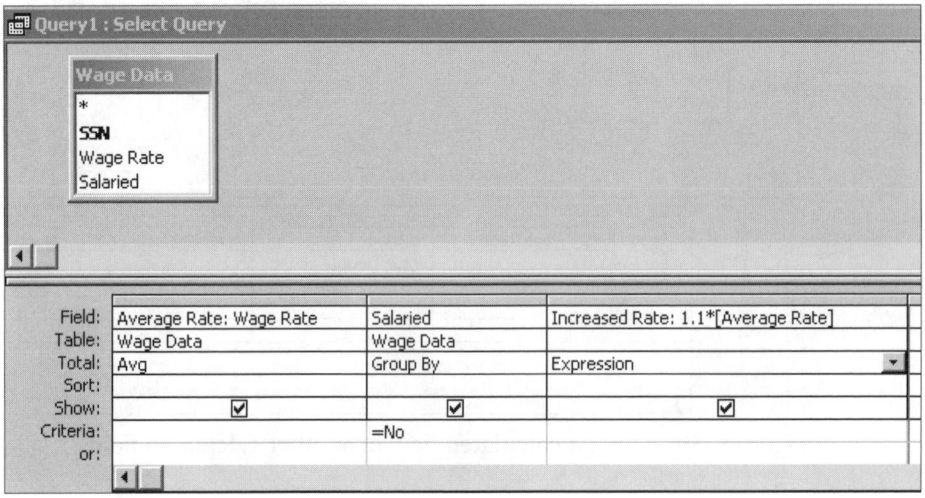

Query1 : Select Query

Wage Data: *, SSN, Wage Rate, Salaried

Field:	Average Rate: Wage Rate	Salaried	Increased Rate: 1.1*[Average Rate]
Table:	Wage Data	Wage Data	
Total:	Avg	Group By	Expression
Sort:			
Show:	☑	☑	☑
Criteria:		=No	
or:			

Figure B-32 An Expression in a Sigma query

Figure B-33 shows the output of the query.

Query1 : Select Query

Average Rate	Salaried	Increased Rate
$10.00	☐	11

Figure B-33 Output of an Expression in a Sigma query

There is no need to save the query definition. Go back to the Design View. Select File—Close.

Using the Date() Function in Queries

Access has two date function features that you should know about. A description of them follows:

1. The following built-in function gives you today's date:

   Date()

   You can use this function in a query criteria or in a calculated field. The function "returns" the day on which the query is run—that is, it puts that value into the place where the function is in an expression.
2. *Date arithmetic* lets you subtract one date from another to obtain the number of days difference. Access would evaluate the following expression as the integer 5 (9 less 4 is 5).

   10/9/2006 – 10/4/2006

Here is an example of how date arithmetic works. Suppose that you want to give each employee a bonus equaling a dollar for each day the employee has worked for you. You'd need to calculate the number of days between the employee's date of hire and the day that the query is run, then multiply that number by 1.

The number of elapsed days is shown by the following equation:

Date() – [Date Hired]

Suppose that for each employee, you want to see the last name, SSN, and bonus amount. You'd set up the query as shown in Figure B-34.

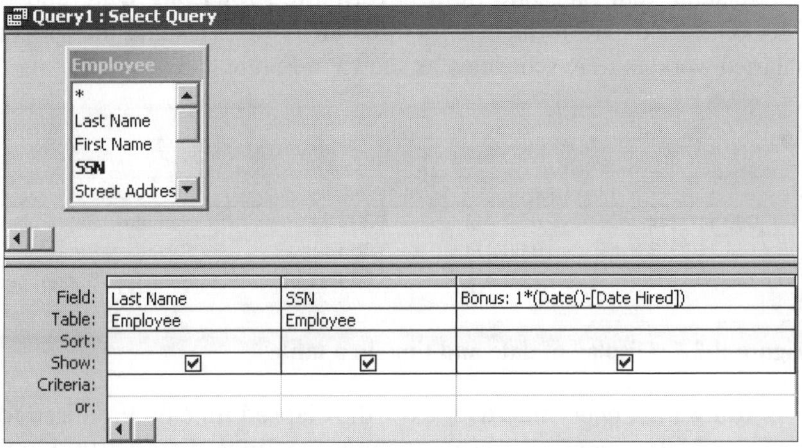

Query1 : Select Query

Employee: *, Last Name, First Name, SSN, Street Addres

Field:	Last Name	SSN	Bonus: 1*(Date()-[Date Hired])
Table:	Employee	Employee	
Sort:			
Show:	☑	☑	☑
Criteria:			
or:			

Figure B-34 Date arithmetic in a query

Assume that you set the format of the Bonus field to Currency. The output will be similar to Figure B-35. (Your Bonus data will be different because you are working on a date different from the date when this tutorial was written.)

Query1 : Select Query

Last Name	SSN	Bonus
Brady	099-11-3344	$0.00
Howard	114-11-2333	$144.00
Smith	123-45-6789	$2,396.00
Smith	148-90-1234	$5,640.00
Jones	222-82-1122	$526.00
Ruth	714-60-1927	$1,226.00

Figure B-35 Output of query with date arithmetic

Using Time Arithmetic in Queries

Access will also let you subtract the values of time fields to get an elapsed time. Assume that your database has a JOB ASSIGNMENTS table showing the times that non-salaried employees were at work during a day. The definition is shown in Figure B-36.

Job Assignments : Table

Field Name	Data Type
SSN	Text
ClockIn	Date/Time
ClockOut	Date/Time
Date	Date/Time

Figure B-36 Date/Time data definition in the JOB ASSIGNMENTS table

Assume that the Date field is formatted for Long Date and that the ClockIn and ClockOut fields are formatted for Medium Time. Assume that, for a particular day, non-salaried workers were scheduled as shown in Figure B-37.

Job Assignments : Table

SSN	ClockIn	ClockOut	Date
099-11-3344	8:30 AM	4:30 PM	Saturday, September 30, 2006
114-11-2333	9:00 AM	3:00 PM	Saturday, September 30, 2006
148-90-1234	7:00 AM	5:00 PM	Saturday, September 30, 2006

Figure B-37 Display of date and time in a table

You want a query that will show the elapsed time on premises for the day. When you add the tables, your screen may show the links differently. Click and drag the JOB ASSIGNMENTS, EMPLOYEE, and WAGE DATA table icons to look like those in Figure B-38.

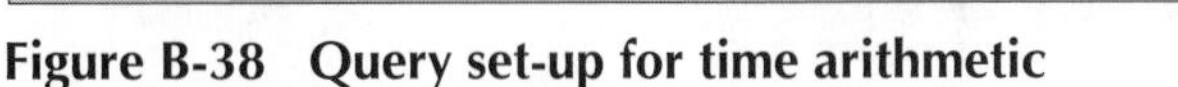

Figure B-38 Query set-up for time arithmetic

Figure B-39 shows the output.

SSN	Salaried	Elapsed Time
114-11-2333	☐	6
148-90-1234	☐	10
099-11-3344	☐	8

Figure B-39 Query output for time arithmetic

The output looks right. For example, employee 099-11-3344 was at work from 8:30 a.m. to 4:30 p.m., which is eight hours. But how does the odd expression that follows yield the correct answers?

([ClockOut] – [ClockIn]) * 24

Why wouldn't the following expression, alone, work?

[ClockOut] – [ClockIn]

This is the answer: In Access, *subtracting one time from the other yields the decimal portion of a 24-hour day*. Employee 099-11-3344 worked 8 hours, which is one-third of a day, so .3333 would result. That is why you must multiply by 24—to convert to an hour basis. Continuing with 099-11-3344, 1/3 x 24 = 8.

Note that parentheses are needed to force Access to do the subtraction *first*, before the multiplication. Without parentheses, multiplication takes precedence over subtraction. With the following expression, ClockIn would be multiplied by 24 and then that value would be subtracted from ClockOut, and the output would be a nonsense decimal number:

[ClockOut] – [ClockIn] * 24

Delete and Update Queries

Thus far, the queries presented in this tutorial have been Select queries. They select certain data from specific tables, based on a given criterion. You can also create queries to update the original data in a database. Businesses do this often, and in real time. For example, when you order an item from a Web site, the company's database is updated to reflect the purchase of the item by deleting it from inventory.

Let's look at an example. Suppose that you want to give all the non-salaried workers a $.50 per hour pay raise. With the three non-salaried workers you have now, it would be easy simply to go into the table and change the Wage Rate data. But assume that you have 3,000 non-salaried employees. It would be much faster and more accurate to change each of the 3,000 non-salaried employees' Wage Rate data by using an Update query to add the $.50 to each employee's wage rate.

AT THE KEYBOARD

Let's change each of the non-salaried employees' pay via an Update query. Figure B-40 shows how to set up the query.

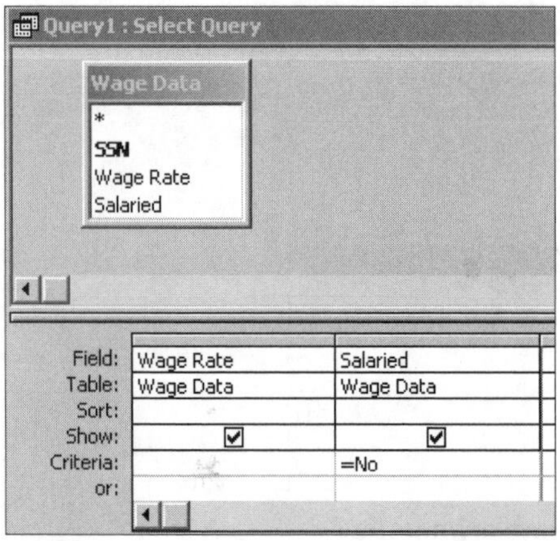

Figure B-40 Query set-up for an Update Query

So far, this query is just a Select query. Place your cursor somewhere above the QBE grid, and then right-click the mouse. Once you are in that menu, choose Query Type—Update Query, as shown in Figure B-41.

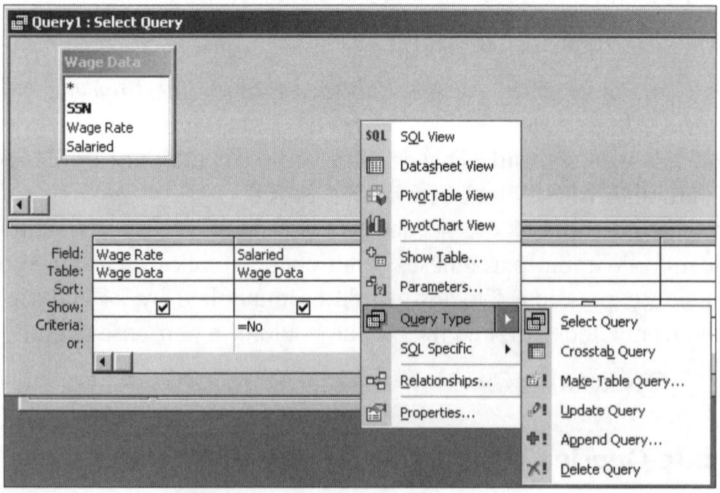

Figure B-41 Selecting a query type

Notice that you now have another line on the QBE grid called "Update to:". This is where you specify the change or update to the data. Notice that you are going to update only the non-salaried workers by using a filter under the Salaried field. Update the Wage Rate data to Wage Rate plus $.50, as shown in Figure B-42. (Note the [ ] as in a calculated field.)

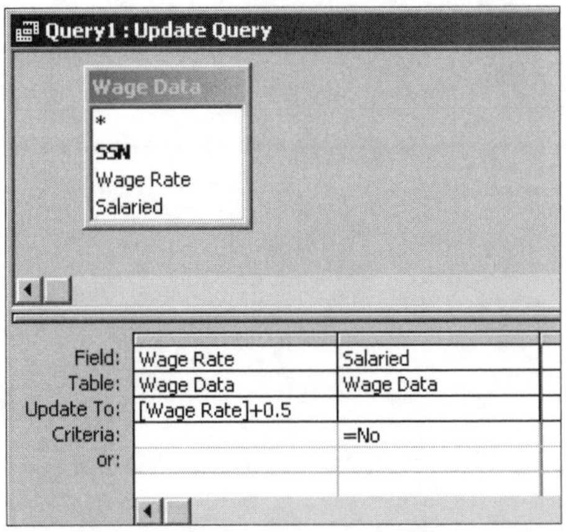

Figure B-42 Updating the wage rate for non-salaried workers

Now run the query. You will first get a warning message, as shown in Figure B-43.

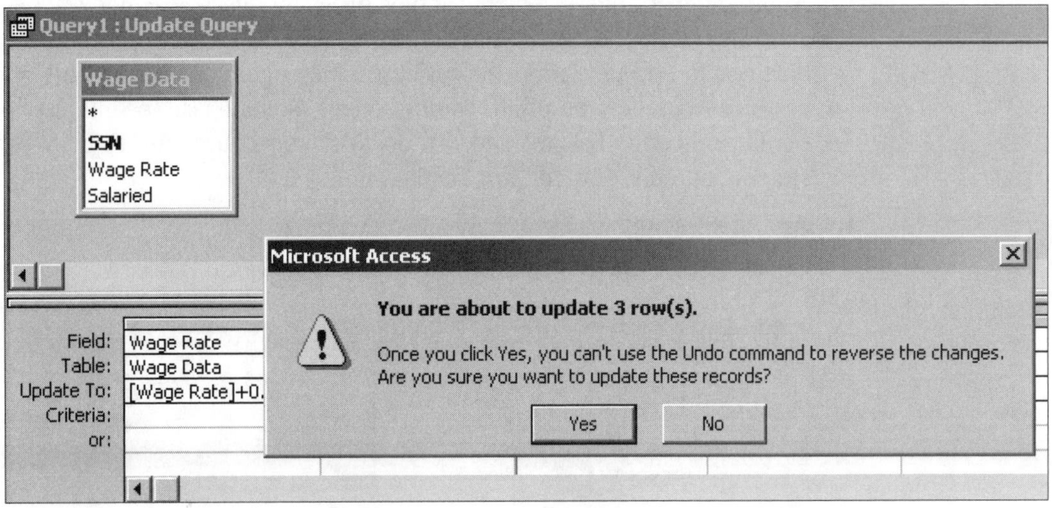

Figure B-43 Update Query warning

Once you click "Yes," the records will be updated. Check those updated records now by viewing the WAGE DATA table. Each salaried wage rate should now be increased by $.50. Note that in this example, you are simply adding $.50 to each salaried wage rate. You could add or subtract data from another table as well. If you do that, remember to call the field name in square brackets.

Delete queries work the same way as Update queries. Assume that your company has been taken over by the state of Delaware. The state has a policy of employing only Delaware residents. Thus, you must delete (or fire) all employees who are not only Delaware residents. To do this, you would first create a Select query using the EMPLOYEE table, right-click your mouse, choose Delete Query from Query Type, then bring down the State field and filter only those records not in Delaware (DE). Do not perform this operation, but note that, if you did, the set-up would look like that in Figure B-44.

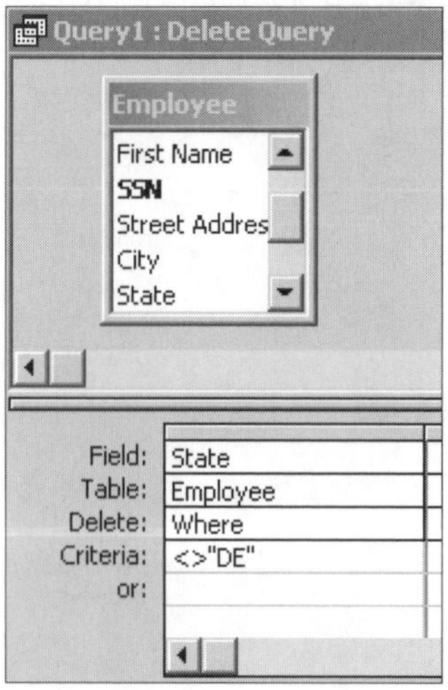

Figure B-44 Deleting all employees who are not Delaware residents

Parameter Queries

Another type of query, which is a type of Select query, is called a **Parameter query**. Here is an example: Suppose that your company has 5,000 employees. You might want to query the database to find the same kind of information again and again, only about different employees. For example, you might want to query the database to find out how many hours a particular employee has worked. To do this, you could run a query previously created and stored, but run it only for a particular employee.

AT THE KEYBOARD

Create a Select query with the format shown in Figure B-45.

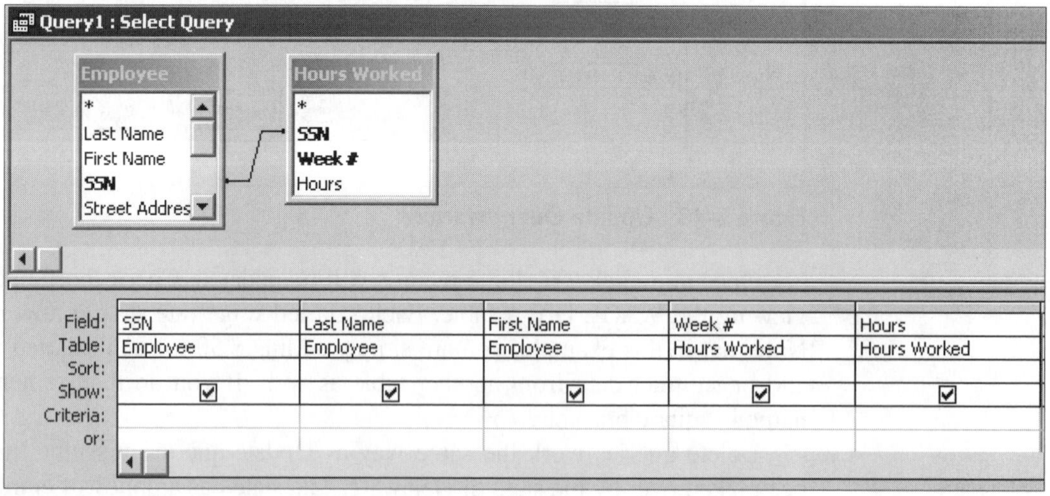

Figure B-45 Design of a Parameter query begins as a Select query

In the Criteria line of the QBE grid for the field SSN, type what is shown in Figure B-46.

Query1 : Select Query

Employee: *, Last Name, First Name, SSN, Street Addres

Hours Worked: *, SSN, Week #, Hours

Field:	SSN	Last Name	First Name	Week #	Hours
Table:	Employee	Employee	Employee	Hours Worked	Hours Worked
Sort:					
Show:	☑	☑	☑	☑	☑
Criteria:	[Enter SSN]				
or:					

Figure B-46 Design of a Parameter query

Note the square brackets, as you would expect to see in a calculated field.

Now run that query. You will be prompted for the specific employee's SSN, as shown in Figure B-47.

Enter Parameter Value

Enter SSN

OK Cancel

Figure B-47 Enter Parameter Value dialog box

Type in your own SSN. Your query output should resemble that shown in Figure B-48.

Query1 : Select Query

SSN	Last Name	First Name	Week #	Hours
099-11-3344	Brady	Joe	1	60
099-11-3344	Brady	Joe	2	55

Figure B-48 Output of a Parameter query

SEVEN PRACTICE QUERIES

This portion of the tutorial is designed to provide you with additional practice in making queries. Before making these queries, you must create the specified tables and enter the records shown in the Creating Tables section of this tutorial. The output shown for the practice queries is based on those inputs.

AT THE KEYBOARD

For each query that follows, you are given a problem statement and a "scratch area." You are also shown what the query output should look like. Follow this procedure: Set up a query in Access. Run the query. When you are satisfied with the results, save the query and continue with the next query. You will be working with the EMPLOYEE, HOURS WORKED, and WAGE DATA tables.

1. Create a query that shows the SSN, last name, state, and date hired for those living in Delaware *and* who were hired after 12/31/92. Sort (ascending) by SSN. (Sorting review: Click in the Sort cell of the field. Choose Ascending or Descending.) Use the table shown in Figure B-49 to work out your QBE grid on paper before creating your query.

Field					
Table					
Sort					
Show					
Criteria					
Or:					

Figure B-49 QBE grid template

Your output should resemble that shown in Figure B-50.

Query 1 : Select Query

SSN	Last Name	State	Date Hired
114-11-2333	Howard	DE	8/1/2006
123-45-6789	Smith	DE	6/1/1996
222-82-1122	Jones	DE	7/15/2004

Figure B-50 Number 1 query output

2. Create a query that shows the last name, first name, date hired, and state for those living in Delaware *or* who were hired after 12/31/92. The primary sort (ascending) is on last name, and the secondary sort (ascending) is on first name. (Review: The Primary Sort field must be to the left of the Secondary Sort field in the query set-up.) Use the table shown in Figure B-51 to work out your QBE grid on paper before creating your query.

Field					
Table					
Sort					
Show					
Criteria					
Or:					

Figure B-51 QBE grid template

If your name were Brady, your output would look like that shown in Figure B-52.

Query 1 : Select Query

Last Name	First Name	Date Hired	State
Brady	Joe	12/23/2002	MN
Howard	Jane	8/1/2006	DE
Jones	Sue	7/15/2004	DE
Ruth	Billy	8/15/1999	MD
Smith	Albert	7/15/1987	DE

Figure B-52 Number 2 query output

Tutorial B

3. Create a query that shows the sum of hours worked by U.S. citizens and by non-U.S. citizens (that is, group on citizenship). The heading for total hours worked should be Total Hours Worked. Use the table shown in Figure B-53 to work out your QBE grid on paper before creating your query.

Field					
Table					
Total					
Sort					
Show					
Criteria					
Or:					

Figure B-53 QBE grid template

Your output should resemble that shown in Figure B-54.

Query1 : Select Query

Total Hours Worked	US Citizen
363	☑
160	☐

Figure B-54 Number 3 query output

4. Create a query that shows the wages owed to hourly workers for Week 1. The heading for the wages owed should be Total Owed. The output headings should be: Last Name, SSN, Week #, and Total Owed. Use the table shown in Figure B-55 to work out your QBE grid on paper before creating your query.

Field					
Table					
Sort					
Show					
Criteria					
Or:					

Figure B-55 QBE grid template

If your name were Joseph Brady, your output would look like that in Figure B-56.

Query1 : Select Query

Last Name	SSN	Week #	Total Owed
Howard	114-11-2333	1	$420.00
Smith	148-90-1234	1	$475.00
Brady	099-11-3344	1	$510.00

Figure B-56 Number 4 query output

5. Create a query that shows the last name, SSN, hours worked, and overtime amount owed for employees paid hourly who earned overtime during Week 2. Overtime is paid at 1.5 times the normal hourly rate for hours over 40. The amount shown should be just the overtime portion of the wages paid. This is not a Sigma query—amounts should be shown for individual workers. Use the table shown in Figure B-57 to work out your QBE grid on paper before creating your query.

Field					
Table					
Sort					
Show					
Criteria					
Or:					

Figure B-57 QBE grid template

If your name were Joseph Brady, your output would look like that shown in Figure B-58.

Query1 : Select Query

Last Name	SSN	Hours	OT Pay
Howard	114-11-2333	50	$157.50
Brady	099-11-3344	55	$191.25

Figure B-58 Number 5 query output

6. Create a Parameter query that shows the hours employees have worked. Have the Parameter query prompt for the week number. The output headings should be Last Name, First Name, Week #, and Hours. Do this only for the non-salaried workers. Use the table shown in Figure B-59 to work out your QBE grid on paper before creating your query.

Field					
Table					
Sort					
Show					
Criteria					
Or:					

Figure B-59 QBE grid template

Run the query with "2" when prompted for the Week #. Your output should look like that shown in Figure B-60.

Query1 : Select Query

Last Name	First Name	Week #	Hours
Howard	Jane	2	50
Smith	Albert	2	40
Brady	Joe	2	55

Figure B-60 Number 6 query output

7. Create an update query that gives certain workers a merit raise. You must first create an additional table as shown in Figure B-61.

Merit Raises : Table

SSN	Merit Raise
114-11-2333	$0.25
148-90-1234	$0.15

Figure B-61 MERIT RAISES table

Now make a query that adds the Merit Raise to the current Wage Rate for those who will receive a raise. When you run the query, you should be prompted with "You are about to update two rows." Check the original WAGE DATA table to confirm the update. Use the table shown in Figure B-62 to work out your QBE grid on paper before creating your query.

Field					
Table					
Update to					
Criteria					
Or:					

Figure B-62 QBE grid template

CREATING REPORTS

Database packages let you make attractive management reports from a table's records or from a query's output. If you are making a report from a table, the Access report generator looks up the data in the table and puts it into report format. If you are making a report from a query's output, Access runs the query in the background (you do not control this or see this happen) and then puts the output in report format.

There are three ways to make a report. One is to handcraft the report in the Design View, from scratch. This is tedious and is not shown in this tutorial. The second way is to use the Report Wizard, during which Access leads you through a menu-driven construction. This method is shown in this tutorial. The third way is to start in the Wizard and then use the Design View to tailor what the Wizard produces. This method is also shown in this tutorial.

Creating a Grouped Report

This tutorial assumes that you can use the Wizard to make a basic ungrouped report. This section of the tutorial teaches you how to make a grouped report. (If you cannot make an ungrouped report, you might learn how to make one by following the first example that follows.)

AT THE KEYBOARD

Suppose that you want to make a report out of the HOURS WORKED table. At the main Objects menu, start a new report by choosing Reports—New. Select the Report Wizard and select the HOURS WORKED table from the drop-down menu as the report basis. Select OK. In the next screen, select all the fields (using the >> button), as shown in Figure B-63.

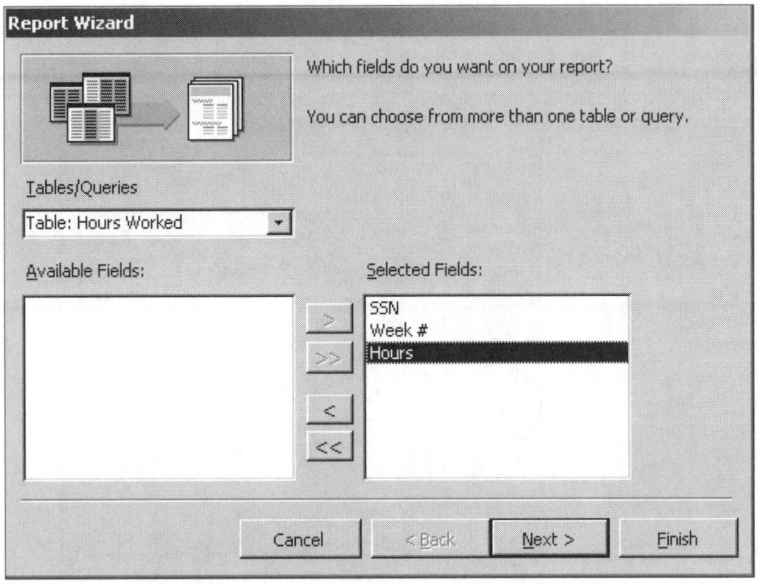

Figure B-63 Field selection step in the Report Wizard

Click Next. Then tell Access that you want to group on Week # by double-clicking that field name. You'll see that shown in Figure B-64.

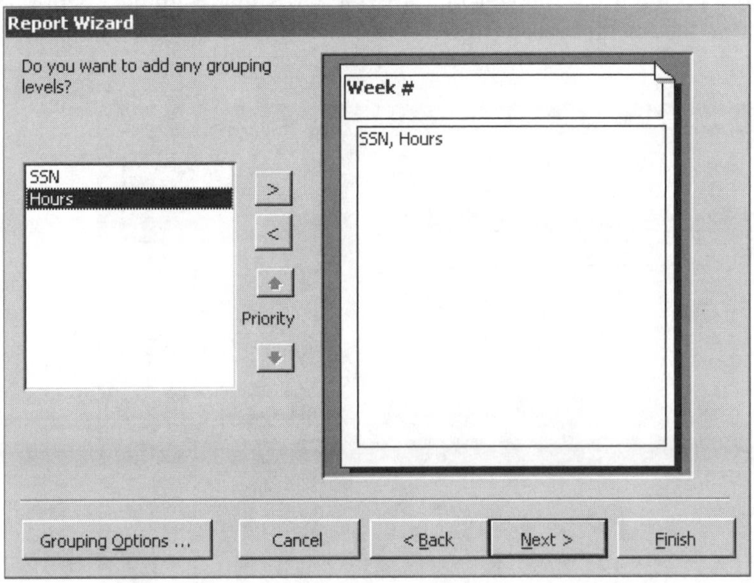

Figure B-64 Grouping step in the Report Wizard

Click Next. You'll see a screen, similar to the one shown in Figure B-65, for Sorting and for Summary Options.

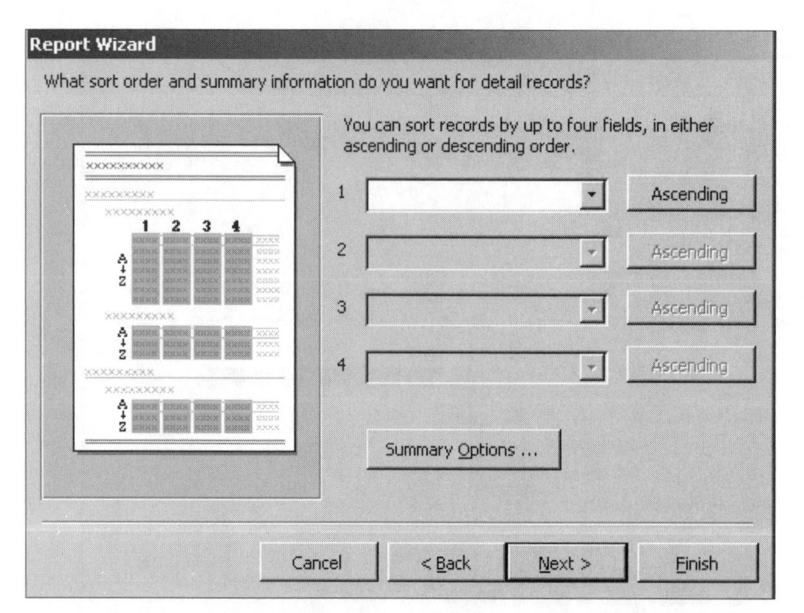

Figure B-65 Sorting and Summary Options step in the Report Wizard

Because you chose a grouping field, Access will now let you decide whether you want to see group subtotals and/or report grand totals. All numeric fields could be added, if you choose that option. In this example, group subtotals are for total hours in each week. Assume that you *do* want the total of hours by week. Click Summary Options. You'll get a screen similar to the one in Figure B-66.

Summary Options
What summary values would you like calculated?
Field Sum Avg Min Max
Hours
OK
Cancel
Show
Detail and Summary
Summary Only
Calculate percent of total for sums

Figure B-66 Summary Options in the Report Wizard

Next, follow these steps:

1. Click the Sum box for Hours (to sum the hours in the group).
2. Click Detail and Summary. (Detail equates with "group," and Summary with "grand total for the report.")
3. Click OK. This takes you back to the Sorting screen, where you can choose an ordering within the group, if desired. (In this case, none is.)

4. Click Next to continue.
5. In the Layout screen (not shown here) choose Stepped and Portrait.
6. Make sure that the "Adjust the field width so all fields fit on a page" check box is unchecked.
7. Click Next.
8. In the Style screen (not shown), accept Corporate.
9. Click Next.
10. Provide a title—Hours Worked by Week would be appropriate.
11. Select the Preview button to view the report.
12. Click Finish.

The top portion of your report will look like that shown in Figure B-67.

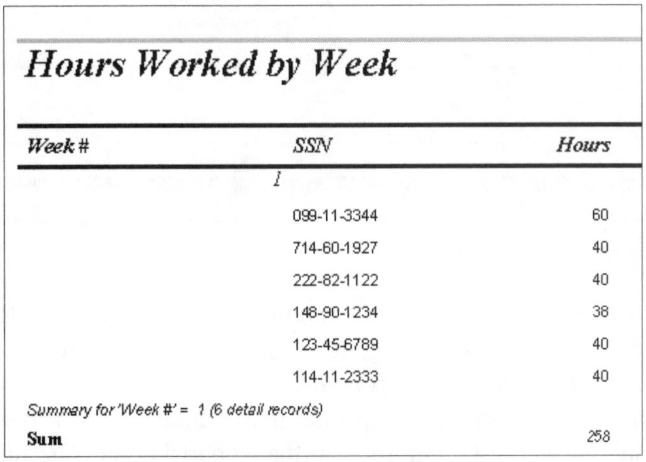

Hours Worked by Week

Week #	*SSN*	*Hours*
1		
	099-11-3344	60
	714-60-1927	40
	222-82-1122	40
	148-90-1234	38
	123-45-6789	40
	114-11-2333	40
Summary for 'Week #' = 1 (6 detail records)		
Sum		258

Figure B-67 Hours Worked by Week report

Notice that data is shown grouped by weeks, with Week 1 on top, then a subtotal for that week. Week 2 data is next, then there is a grand total (which you can scroll down to see). The subtotal is labeled "Sum," which is not very descriptive. This can be changed later in the Design View. Also, there is the apparently useless italicized line that starts out *"Summary for 'Week ..."* This also can be deleted later in the Design View. At this point, you should select File—Save As (accept the suggested title if you like). Then select File—Close to get back the Database window. Try it. Your report's Objects screen should resemble that shown in Figure B-68.

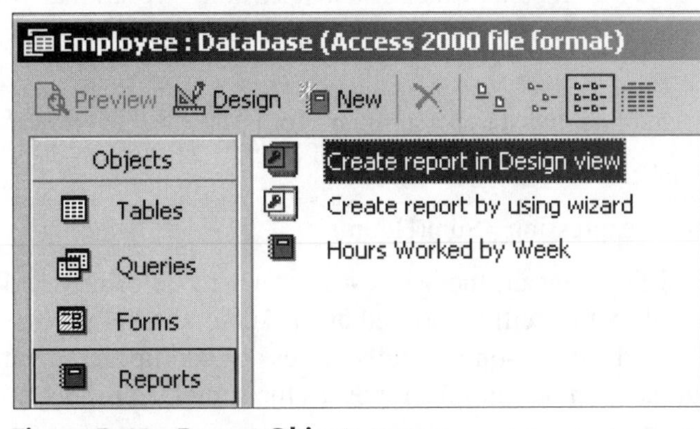

Figure B-68 Report Objects screen

To edit the report in the Design View, click the report title, then the Design button. You will see a complex (and intimidating) screen, similar to the one shown in Figure B-69.

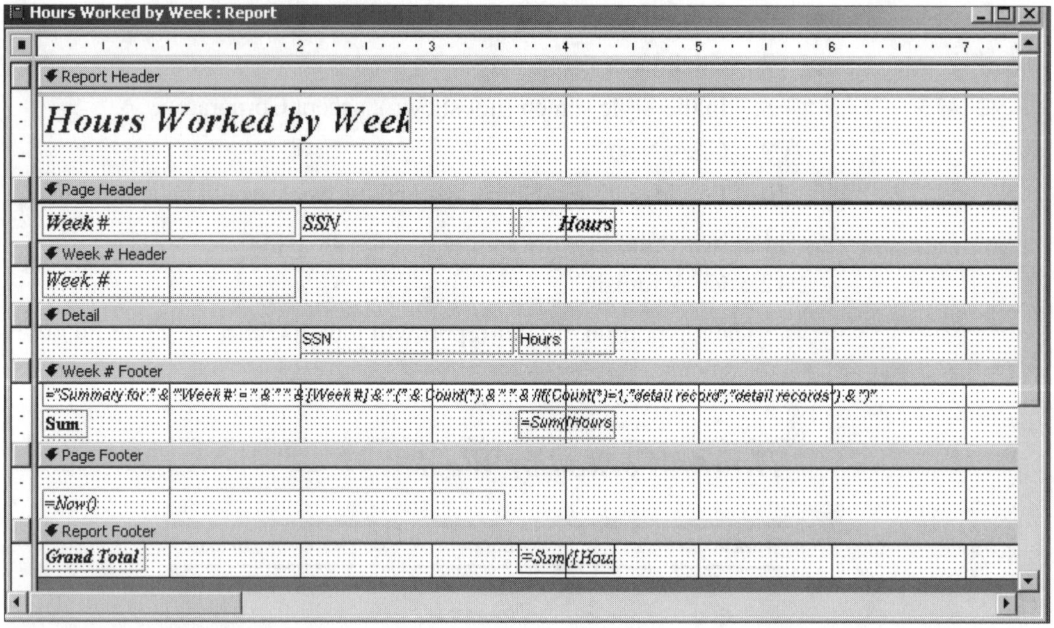

Figure B-69 Report design screen

The organization of the screen is hierarchical. At the top is the Report level. The next level down (within a report) is the Page level. The next level or levels down (within a page) are for any data groupings you have specified.

If you told Access to make group (summary) totals, your report will have a Report Header area and end with a Grand Total in the Report Footer. The report header is usually just the title you have specified.

A page also has a header, which is usually just the names of the fields you have told Access to put in the report (here, Week #, SSN, and Hours fields). Sometimes the page number is put in by default.

Groupings of data are more complex. There is a header for the group—in this case, the *value* of the Week # will be the header; for example, there is a group of data for the first week, then one for the second—the values shown will be 1 and 2. Within each data grouping is the other "detail" that you've requested. In this case, there will be data for each SSN and the related hours.

Each Week # gets a "footer," which is a labeled sum—recall that you asked for that to be shown (Detail and Summary were requested). The Week # Footer is indicated by three things:

1. The italicized line that starts =Summary for ...
2. The Sum label
3. The adjacent expression =Sum(Hours)

The italicized line beneath the Week # Footer will be printed unless you eliminate it. Similarly, the word "Sum" will be printed as the subtotal label unless you eliminate it. The "=Sum(Hours)" is an expression that tells Access to add up the quantity *for the header in question* and put that number into the report as the subtotal. (In this example, that would be the sum of hours, by Week #.)

Each report also gets a footer—the grand total (in this case, of hours) for the report.

If you look closely, each of the detail items appears to be doubly inserted in the design. For example, you will see the notation for SSN twice, once in the Page Header and then again in the Detail band. Hours are treated similarly.

The data items will not actually be printed twice, because each data element is an object in the report; each object is denoted by a label and by its value. There is a representation of the name, which is the boldface name itself (in this example, "SSN" in the page header), and there is a representation in less-bold type for the value "SSN" in the Detail band.

Sometimes, the Report Wizard is arbitrary about where it puts labels and data. If you do not like where the Wizard puts data, the objects containing data can be moved around in the Design View. You can click and drag within the band or across bands. Often, a box will be too small to allow full numerical values to show. When that happens, select the box and then click one of the sides to stretch it. This will allow full values to show. At other times, an object's box will be very long. When that happens, the box can be clicked, re-sized, then dragged right or left in its panel to reposition the output.

Suppose that you do *not* want the italicized line to appear in the report. Also suppose that you would like different subtotal and grand total labels. The italicized line is an object that can be activated by clicking it. Do that. "Handles" (little squares) appear around its edges, as shown in Figure B-70.

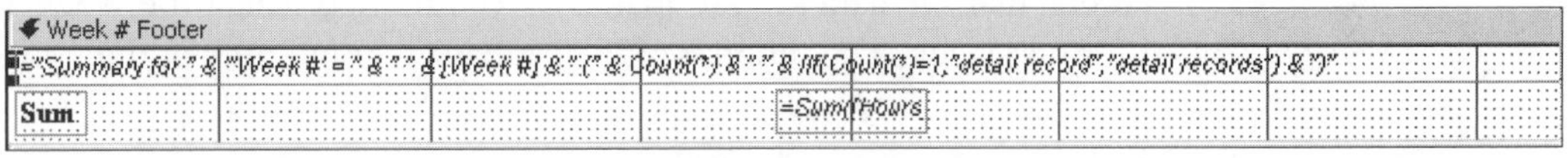

Figure B-70 Selecting an object in the Report Design View

Press the Delete key to get rid of the selected object.

To change the subtotal heading, click the Sum object, as shown in Figure B-71.

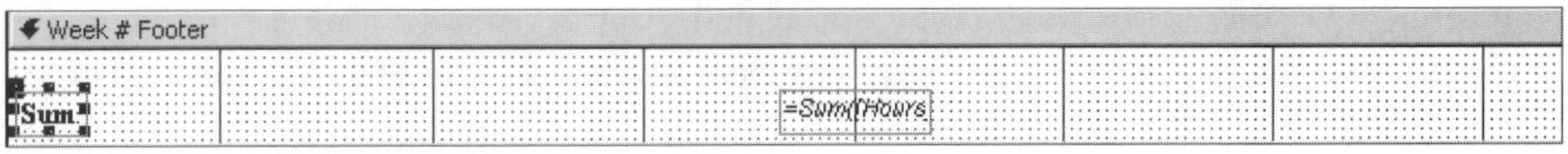

Figure B-71 Selecting the Sum object in the Report Design View

Click again. This gives you an insertion point from which you can type, as shown in Figure B-72.

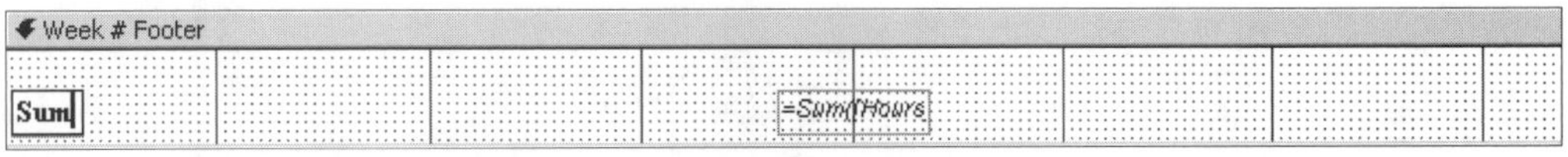

Figure B-72 Typing in an object in the Report Design View

Change the label to something like Sum of Hours for Week, then hit Enter, or click somewhere else in the report to deactivate. Your screen should resemble that shown in Figure B-73.

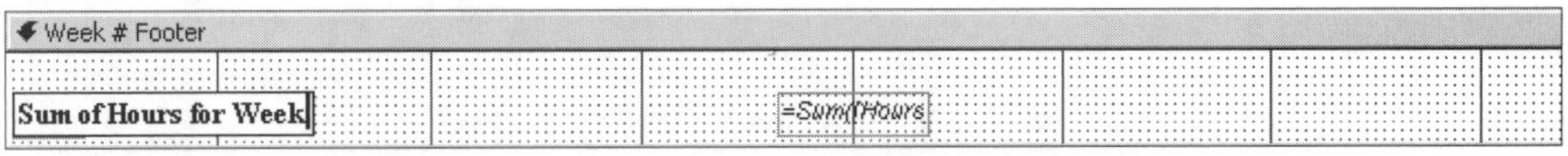

Figure B-73 Changing a label in the Report Design View

You can change the Grand Total in the same way.

Finally, you'll want to save and then print the file: Select File—Save, then select File—Print Preview. You should see a report similar to that in Figure B-74 (top part is shown).

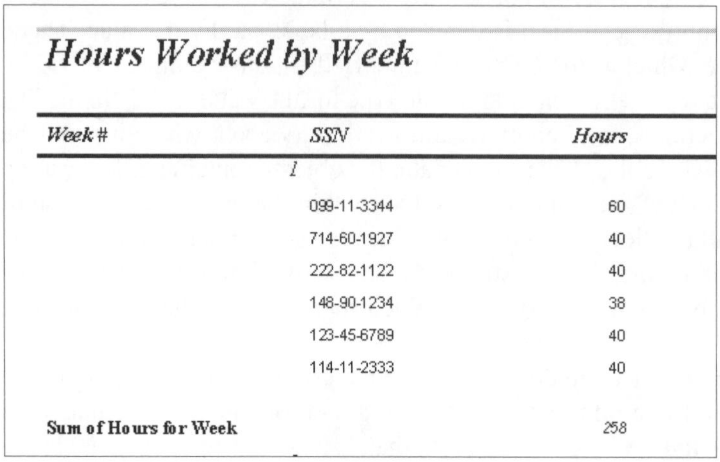

Hours Worked by Week

Week #	*SSN*	*Hours*
	1	
	099-11-3344	60
	714-60-1927	40
	222-82-1122	40
	148-90-1234	38
	123-45-6789	40
	114-11-2333	40
Sum of Hours for Week		*258*

Figure B-74 Hours Worked by Week report

Notice that the data are grouped by week number (data for Week 1 is shown) and subtotaled for that week. The report would also have a grand total at the bottom.

Moving Fields in the Design View

When you group on more than one field in the Report Wizard, the report has an odd "staircase" look. There is a way to overcome that effect in the Design View, which you will learn next.

Suppose that you make a query showing an employee's last name, street address, zip code, and wage rate. Then you make a report from that query, grouping on last name, street address, and zip code. (Why you would want to organize a report in this way is not clear, but for the moment, accept the organization for the purpose of the example.) This is shown in Figure B-75.

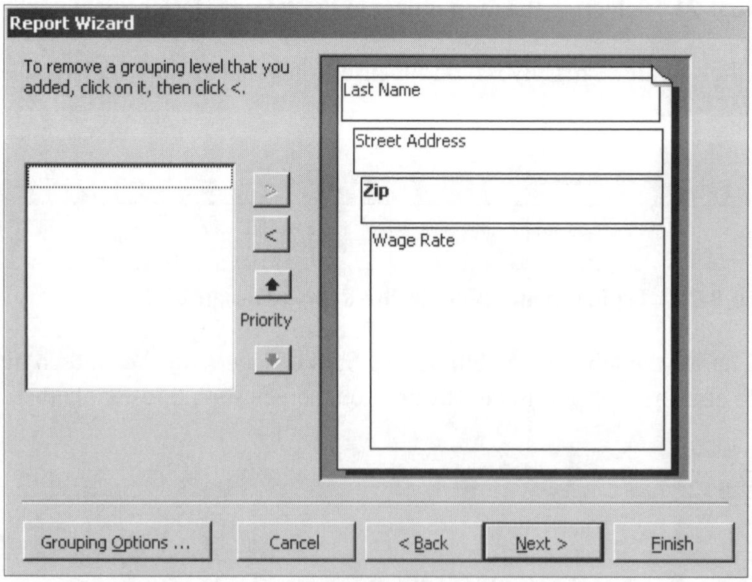

Figure B-75 Grouping in the Report Wizard

Then, follow these steps:

1. Click Next.
2. You do not Sum anything in Summary Options.
3. Click off the check mark by “Adjust the field width so all fields fit on a page”.
4. Select Landscape.
5. Select Stepped. Click Next.
6. Select Corporate. Click Next.
7. Type a title (Wage Rates for Employees). Click Finish.

When you run the report, it will have a “staircase” grouped organization. In the report that follows in Figure B-76, notice that Zip data is shown below Street Address data, and Street Address data is shown below Last Name data. (The field Wage Rate is shown subordinate to all others, as desired. Wage rates may not show on the screen without scrolling.)

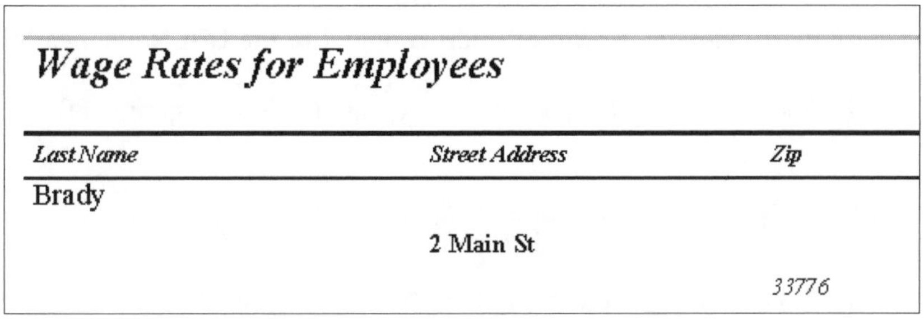

Figure B-76 Wage Rates for Employees grouped report (Wage Rate not shown)

Suppose that you want the last name, street address, and zip all on the same line. The way to do that is to take the report into the Design View for editing. At the Database window, select “Wage Rates for Employees” Report and Design. At this point, the headers look like those shown in Figure B-77.

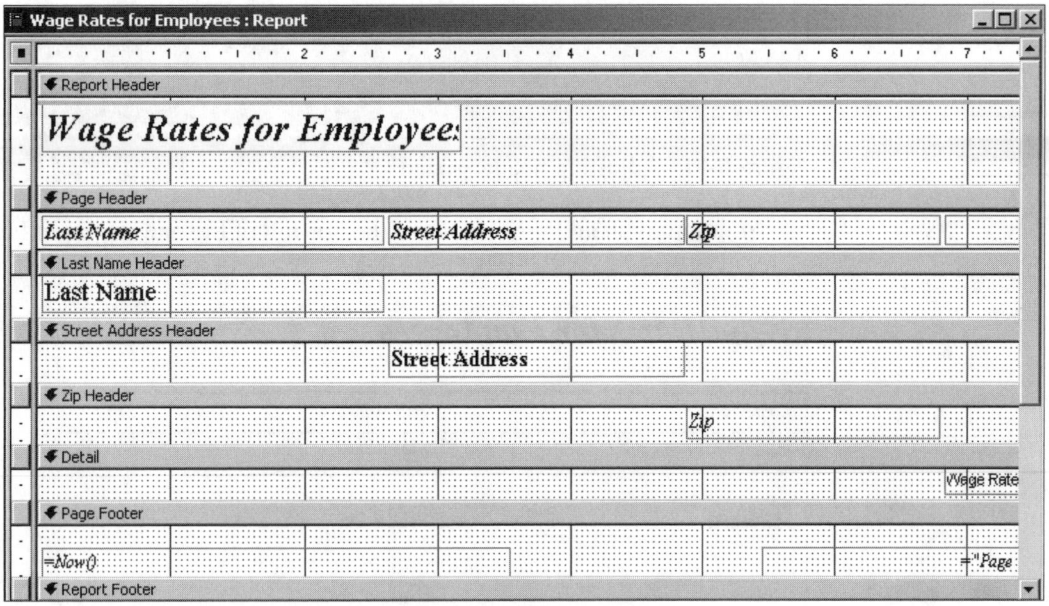

Figure B-77 Wage Rates for Employees report Design View

Your goal is to get the Street Address and Zip fields into the last name header (*not* into the page header!), so they will then print on the same line. The first step is to click the Street Address object in the Street Address Header, as shown in Figure B-78.

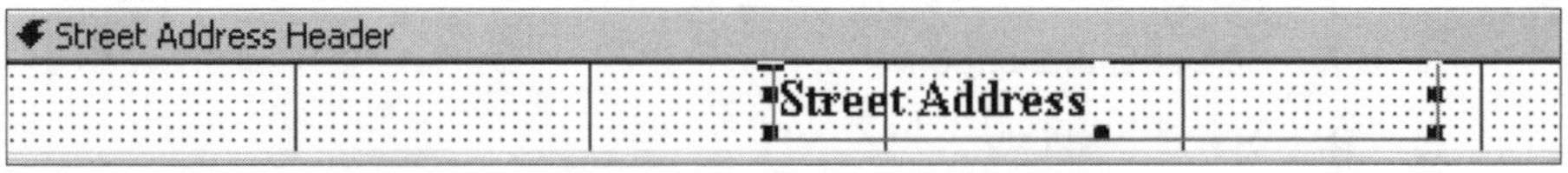

Figure B-78 Selecting Street Address object in the Street Address header

Hold down the button with the little hand icon, and drag the object up into the Last Name Header, as shown in Figure B-79.

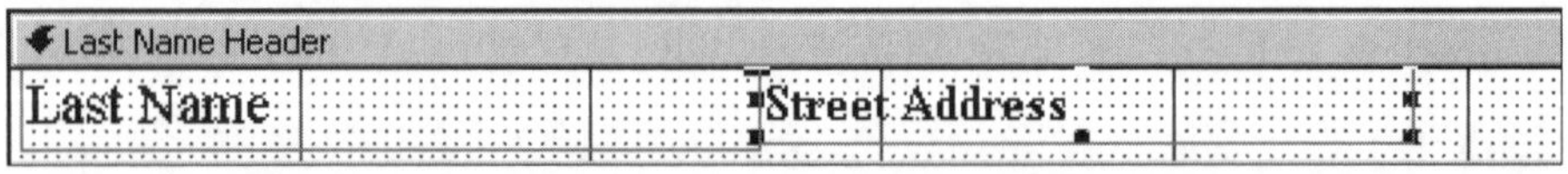

Figure B-79 Moving the Street Address object to the Last Name header

Do the same thing with the Zip object, as shown in Figure B-80.

Figure B-80 Moving the Zip object to the Last Name header

To get rid of the header space allocated to the objects, tighten up the "dotted" area between each header. Put the cursor on the top of the header panel. The arrow changes to something that looks like a crossbar. Click and drag it up to close the distance. After both headers are moved up, your screen should look like that shown in Figure B-81.

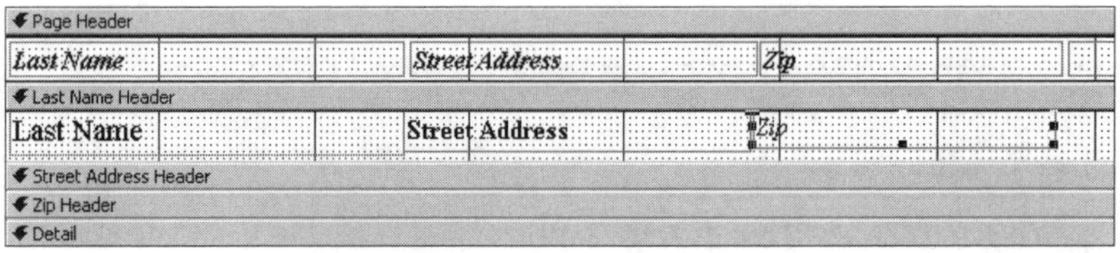

Figure B-81 Adjusting header space

Your report should now resemble the portion of the one shown in Figure B-82.

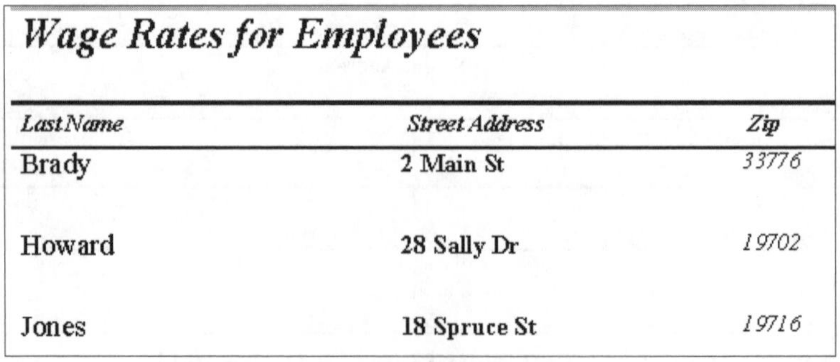

Wage Rates for Employees

LastName	Street Address	Zip
Brady	2 Main St	33776
Howard	28 Sally Dr	19702
Jones	18 Spruce St	19716

Figure B-82 Wage Rates for Employees report

IMPORTING DATA

Text or spreadsheet data is easily imported into Access. In business, importing data happens frequently due to disparate systems. Assume that your healthcare coverage data is on the Human Resources Manager's computer in an Excel spreadsheet. Open the software application Microsoft Excel. Create that spreadsheet in Excel now, using the data shown in Figure B-83.

	A	B	C
1	SSN	Provider	Level
2	114-11-2333	BlueCross	family
3	123-45-6789	BlueCross	family
4	148-90-1234	Coventry	spouse
5	222-82-1122	None	none
6	714-60-1927	Coventry	single
7	Your SSN	BlueCross	single

Figure B-83 Excel data

Save the file, then close it. Now you can easily import that spreadsheet data into a new table in Access. With your **Employee** database open and Tables object selected, click New and click Import Table, as shown in Figure B-84. Click OK.

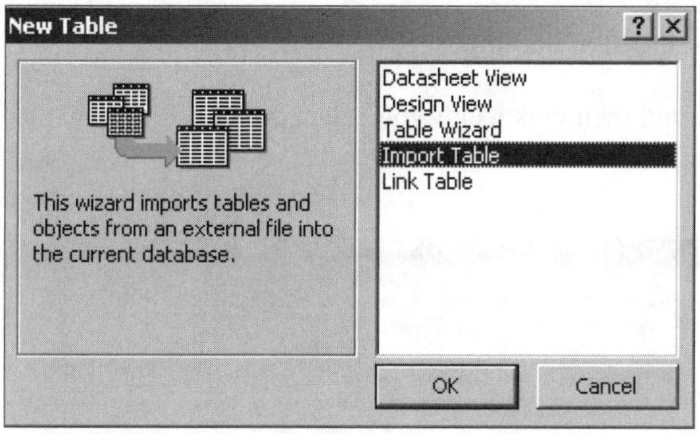

New Table

This wizard imports tables and objects from an external file into the current database.

Datasheet View
Design View
Table Wizard
Import Table
Link Table

OK Cancel

Figure B-84 Importing data into a new table

Find and import your spreadsheet. Be sure to choose **Microsoft Excel** as **Files of Type**. Assuming that you just have one worksheet in your Excel file, your next screen looks like that shown in Figure B-85.

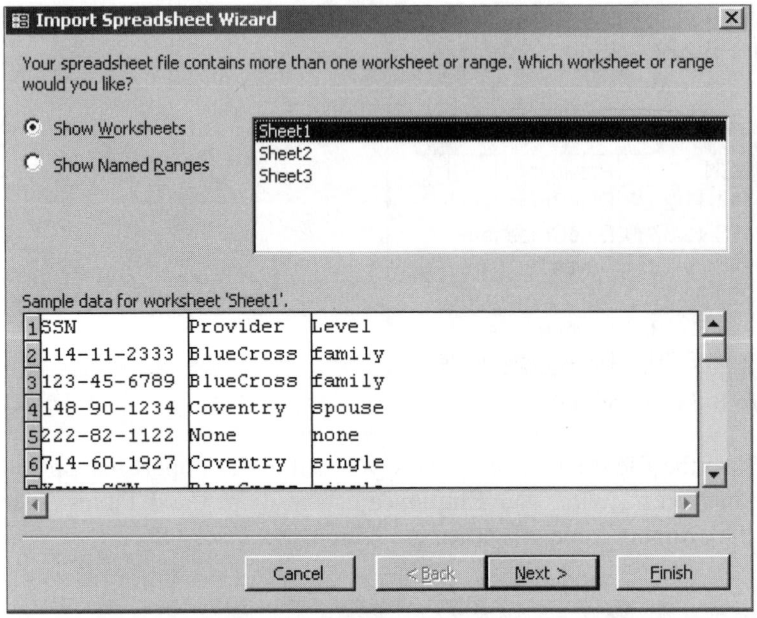

Figure B-85 First screen in the Import Spreadsheet Wizard

Choose Next, and then make sure you select the check box that says First Row Contains Column Headings, as shown in Figure B-86.

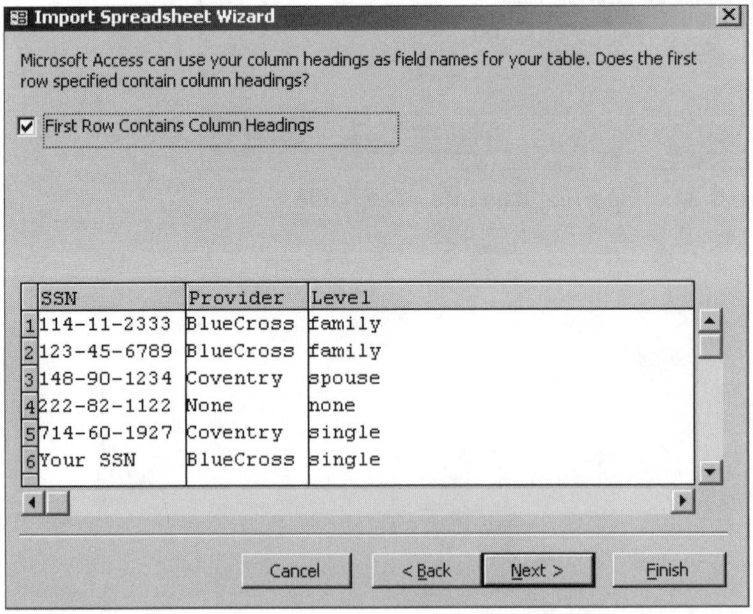

Figure B-86 Choosing column headings in the Import Spreadsheet Wizard

Store your data in a new table, do not index anything (you'll see this in the next screen of the Wizard), but choose your own primary key, which would be SSN, as chosen in Figure B-87.

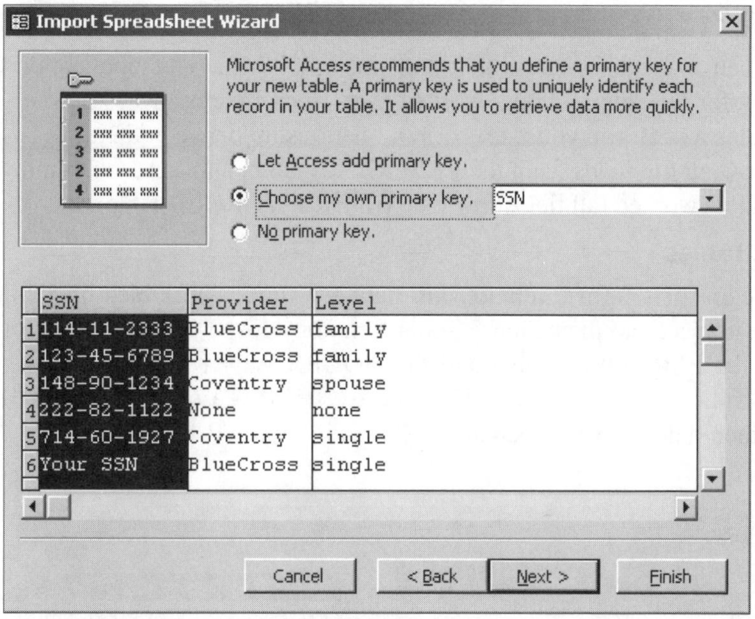

Import Spreadsheet Wizard

Microsoft Access recommends that you define a primary key for your new table. A primary key is used to uniquely identify each record in your table. It allows you to retrieve data more quickly.

Let Access add primary key.

Choose my own primary key. SSN

No primary key.

SSN	Provider	Level
114-11-2333	BlueCross	family
123-45-6789	BlueCross	family
148-90-1234	Coventry	spouse
222-82-1122	None	none
714-60-1927	Coventry	single
Your SSN	BlueCross	single

Cancel < Back Next > Finish

Figure B-87 Choosing a primary key field in the Import Spreadsheet Wizard

Continue through the Wizard, giving your table an appropriate name. After the table is imported, take a look at it and its design. (Highlight the Table option and use the Design button.) Note the width of each field (very large). Adjust the field properties as needed.

FORMS

Forms simplify adding new records to a table. The Form Wizard is easy to use and can be performed on a single table or on multiple tables.

When you base a form on one table, you simply identify that table when you are in the Form Wizard set-up. The form will have all the fields from that table and only those fields. When data is entered into the form, a complete new record is automatically added to the table.

But what if you need a form that includes the data from two (or more) tables? Begin (counterintuitively) with a query. Bring all tables you need in the form into the query. Bring down the fields you need from each table. (For data-entry purposes, this probably means bringing down *all* the fields from each table.) All you are doing is selecting fields that you want to show up in the form, so you make *no criteria* after bringing fields down in the query. Save the query. When making the form, tell Access to base the form on the query. The form will show all the fields in the query; thus, you can enter data into all the tables at once.

Suppose that you want to make one form that would, at the same time, enter records into the EMPLOYEE table and the WAGE DATA table. The first table holds relatively permanent data about an employee. The second table holds data about the employee's starting wage rate, which will probably change.

The first step is to make a query based on both tables. Bring down all the fields from both tables into the lower area. Basically, the query just gathers up all the fields from both tables into one place. No criteria are needed. Save the query.

The second step is to make a form based on the query. This works because the query knows about all the fields. Tell the form to display all fields in the query. (Common fields—here, SSN—would appear twice, once for each table.)

Forms with Subforms

You can also make a form that contains a subform. This application would be particularly handy for viewing all hours worked each week by employee. Before you create a form that contains a subform, you must form a relationship between the tables. Suppose that you want to show all the fields from the EMPLOYEE table, and for each employee, you want to show the hours worked (all fields from the HOURS WORKED table).

Join the Tables

To begin, first form a relationship between those two tables by joining them: Choose the Tables object and then choose Tools-Relationships. The Show Table dialog box will pop up. Add the EMPLOYEE table and the HOURS WORKED table. Drag your cursor from the SSN field in the EMPLOYEE table to the SSN field in the HOURS WORKED table. Another dialog box will pop up, as shown in Figure B-88.

Figure B-88 The Edit Relationships dialog box

Click the Join Type button, and choose Number 2: *Include ALL records from 'Employee' and only those records from 'Hours Worked' where the joined fields are equal,* as shown in Figure B-89.

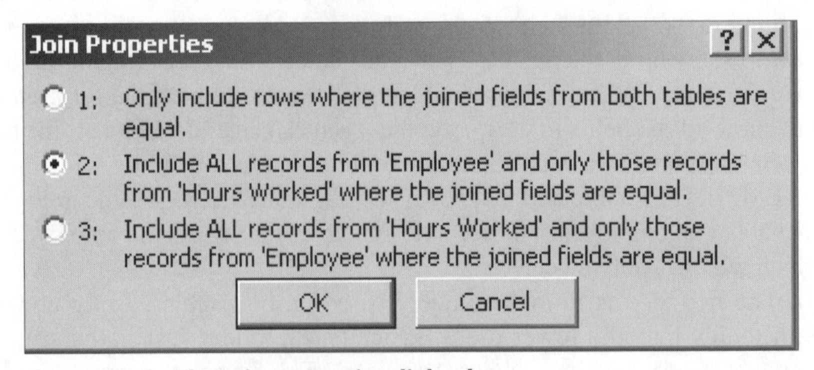

Figure B-89 The Join Properties dialog box

Click OK, then click Create. Close the Edit Relationships window and save the changes.

Create the Form and Subform

To create the form and subform, first create a simple, one-table form using the Form Wizard on the EMPLOYEE table. Follow these steps:

1. In the Forms Object, choose Create form **by using Wizard**.
2. Make sure the table Employee is selected under the drop-down menu of Tables/Queries.
3. Select all Available Fields by clicking the right double-arrow button.
4. Select Next.
5. Select Columnar layout.
6. Select Next.
7. Select Standard Style.
8. Select Next.
9. When asked, "What title do you want for your form?", type Employee Hours.
10. Select Finish.

After the form is complete, click on the Design View, so your screen looks like the one shown in Figure B-90.

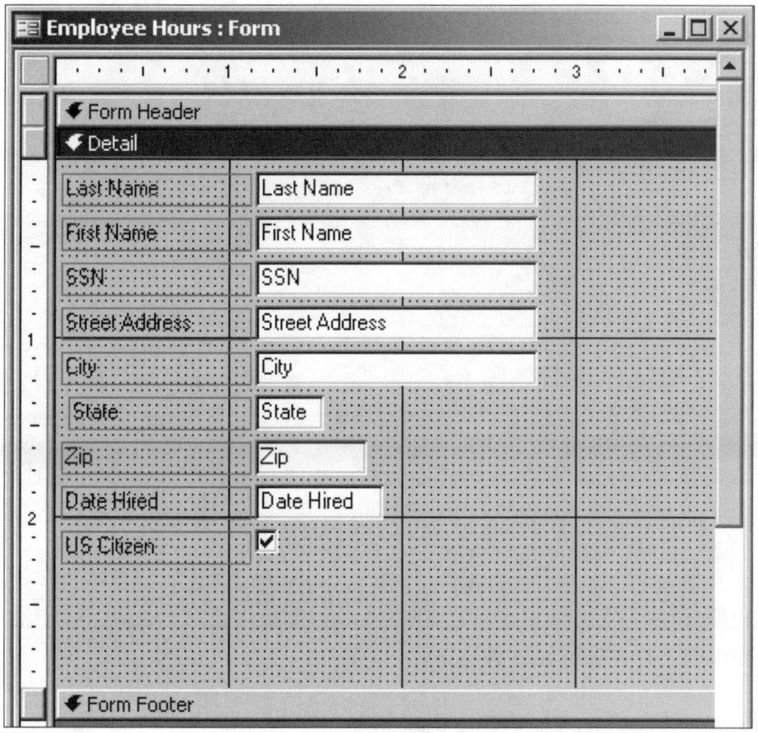

Figure B-90 The Employee Hours form

Make sure the Toolbox window is showing on the screen (Figure B-91). If it is not visible, select View—Toolbox. (The Toolbox may also appear as a toolbar for some students.)

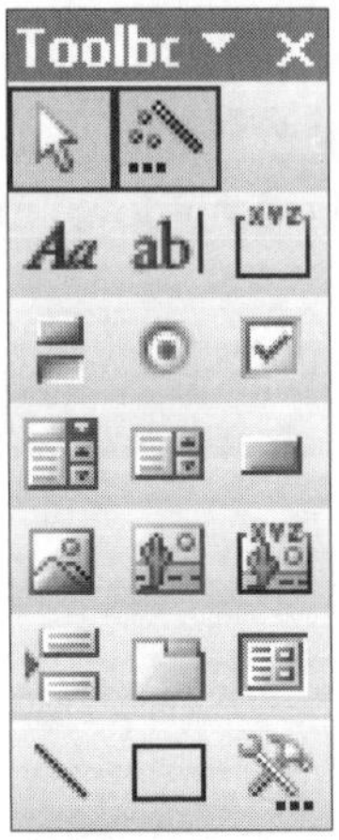

Figure B-91 The Toolbox window

Click the Subform/Subreport button (6th row, button on right) and, using your cursor, drag a small section next to the State, Zip, Date Hired, and US Citizen fields in your form design. As you lift your cursor, the Subform Wizard will appear, as shown in Figure B-92.

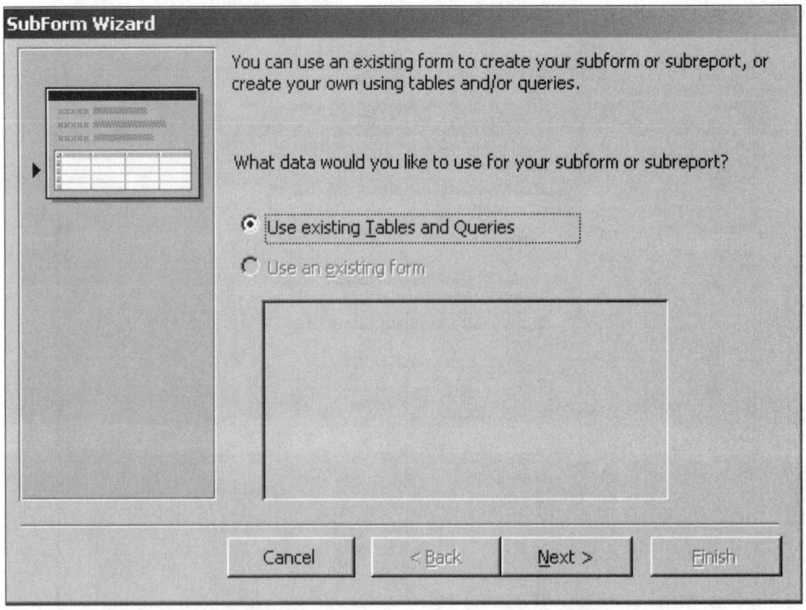

Figure B-92 The Subform Wizard

Follow these steps to create data in the subform:

1. Select the button Use Existing Tables and Queries.
2. Select Next.
3. Under Tables/Queries, choose the HOURS WORKED table, and bring all fields into the Selected Fields box by clicking the right double-arrow button.
4. Select Next.

5. Select the Choose **from a list** radio button.
6. Select Next.
7. Use the default subform name.
8. Select Finish.

Now you will need to adjust the design so all fields' data are visible. Go to the Datasheet View, and click through the various records to see how the subform data changes. Your final form should resemble the one shown in Figure B-93.

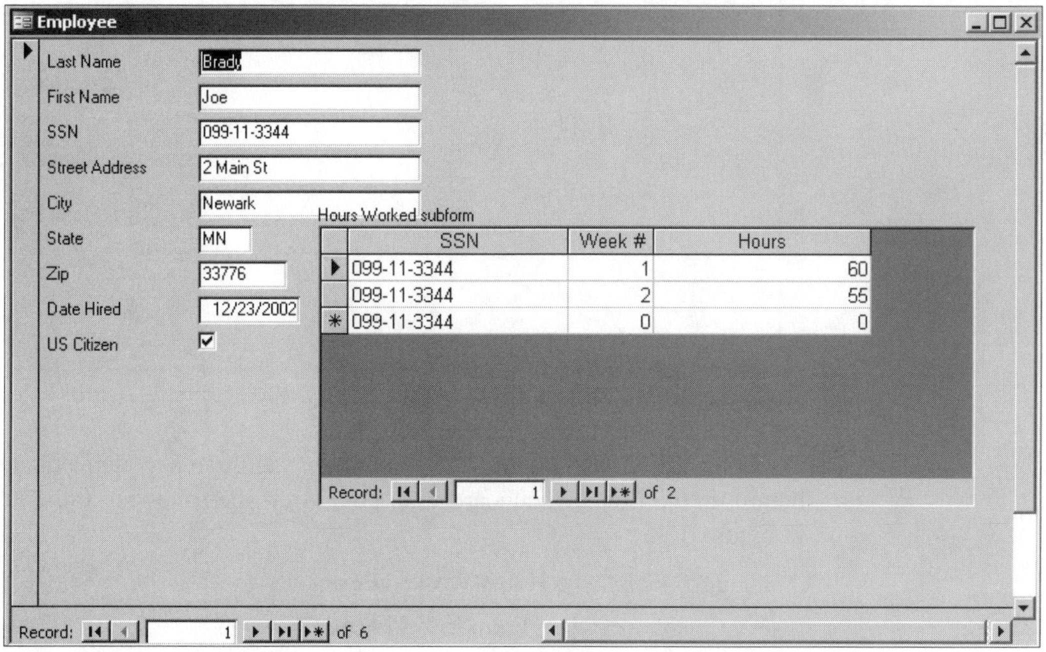

Figure B-93 The Employee Hours form with the Hours Worked subform

Create a Switchboard Form

If you want someone who knows nothing about Access to run your Access database, you can use the Switchboard Manager to create a Switchboard form to simplify their work. A Switchboard form provides a simple, user-friendly interface that has buttons to click to do certain tasks. For example, you could design a Switchboard with three buttons: one for the Employee Hours Worked form, one for the Wage Rates for Employees report, and one for the Hours Worked by Week report. Your finished product will be a page showing three buttons. Each button can be clicked to open either the form, or one of the two reports. To design that Switchboard, use the following steps:

1. Remain on the Forms Object.
2. Select Tools.
3. Select Database Utilities.
4. Select Switchboard Manager.
5. A screen will prompt you with the question, "The Switchboard Manager was unable to find a valid switchboard in this database. Would you like to create one?" Click Yes.

The Switchboard Manager screen will open, as shown in Figure B-94. Leaving the Switchboard (Default) highlighted, click the Edit button.

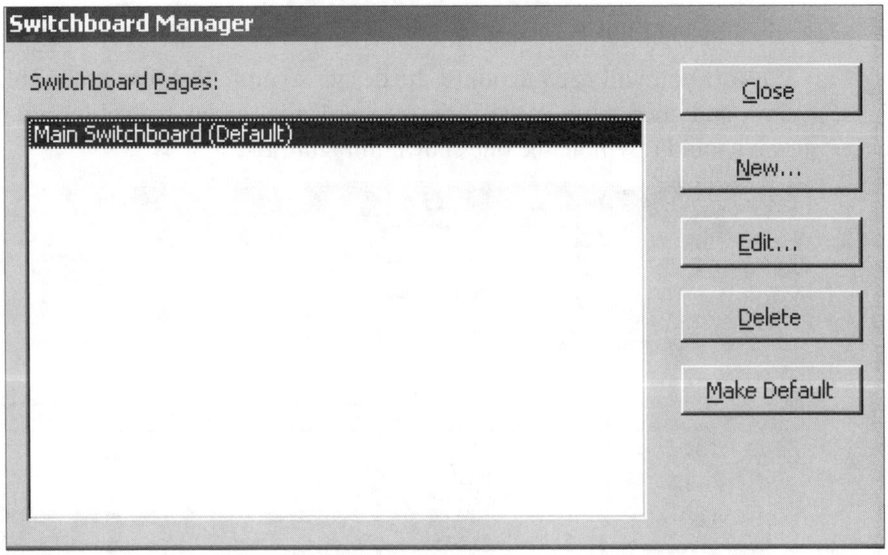

Figure B-94 The Switchboard Manager screen

In the Edit Switchboard page, you will create three new items on the page. Click the New button. In the Edit Switchboard Item box, insert the following three items of data (as shown in Figure B-95):

1. *Text:* Employee Hours Worked Form
2. *Command:* Open Form in Add Mode
3. *Form:* Employee Hours

Click OK when you are finished.

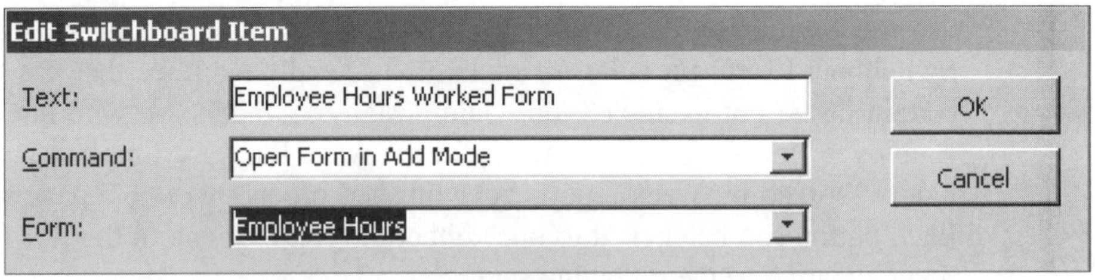

Figure B-95 The Edit Switchboard Item screen

You will repeat this procedure two more times (that is, click the New button in the Edit Switchboard Page). Next, insert the following data:

1. *Text:* Wage Rate for Employees Report
2. *Command:* Open Report
3. *Report:* Wage Rate for Employees

Click OK when you are finished. Then, repeat the procedure (that is, click the New button in the Edit Switchboard Page) and insert the following data:

1. *Text:* Hours Worked by Week Report

2. *Command:* Open Report
3. *Report:* Hours Worked by Week

Click OK when you are finished. At this point, your Edit Switchboard screen should look like Figure B-96.

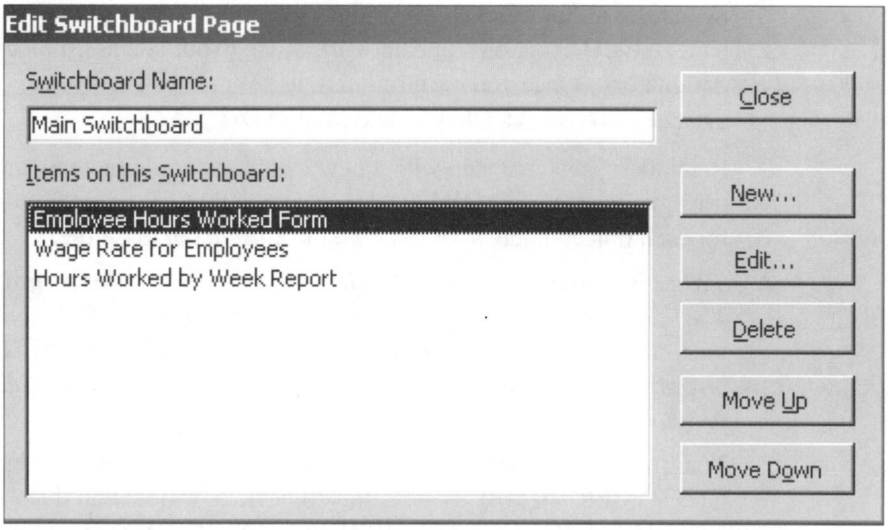

Edit Switchboard Page

Switchboard Name:

Main Switchboard

Items on this Switchboard:

Employee Hours Worked Form
Wage Rate for Employees
Hours Worked by Week Report

Close | New... | Edit... | Delete | Move Up | Move Down

Figure B-96 The Edit Switchboard Page

Click the Close button, and then click the Close button again.

You can test the Switchboard by clicking the Switchboard in the Forms Objects. It should look like that shown in Figure B-97.

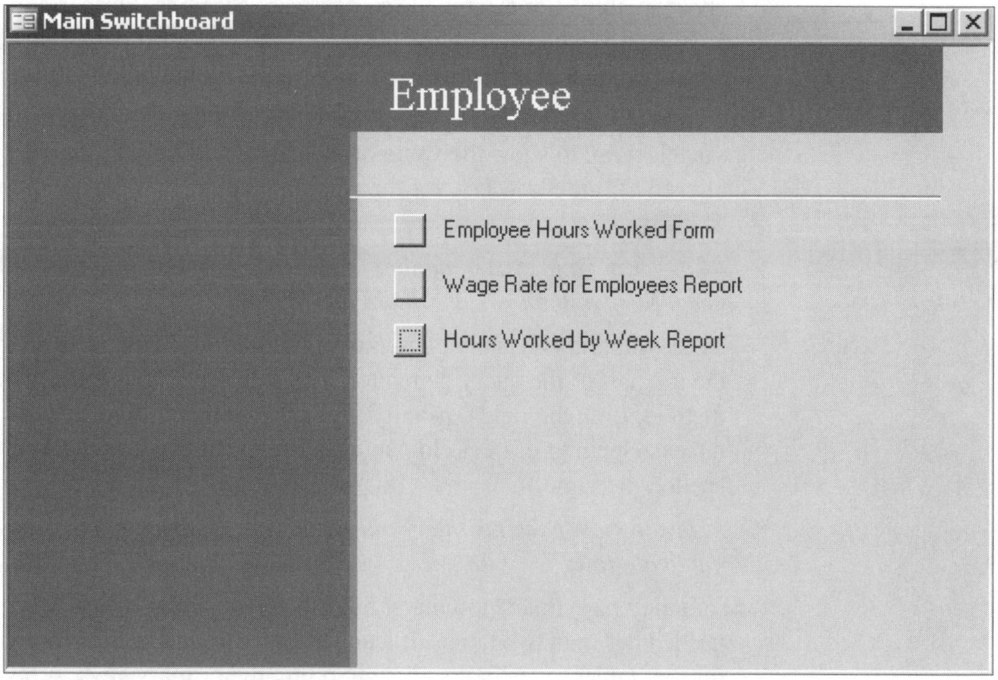

Main Switchboard

Employee

Employee Hours Worked Form
Wage Rate for Employees Report
Hours Worked by Week Report

Figure B-97 The Main Switchboard showing one form and two reports

Troubleshooting Common Problems

Access beginners (and veterans!) sometimes create databases that have problems. Common problems are described here, along with their causes and corrections.

1. *"I saved my database file, but it is not on my disk! Where is it?"*

   You saved to some fixed disk. Use the Search option of the Windows Start button. Search for all files ending in ".mdb" (search for *.mdb). If you did save it, it is on the hard **Drive (C:\)** or on some network drive. (Your site assistant can tell you the drive designators.) Once you have found it, use Windows Explorer to copy it to your diskette in **Drive A:**. Click it, and drag to **Drive A:**.

   Reminder: Your first step with a new database should be to Open it on the intended drive, which is usually **Drive A:** for a student. Don't rush this step. Get it right. Then, for each object made, save it *within* the current database file.

2. *"What is a 'duplicate key field value'? I'm trying to enter records into my Sales table. The first record was for a sale of product X to customer #101, and I was able to enter that one. But when I try to enter a second sale for customer #101, Access tells me I already have a record with that key field value. Am I only allowed to enter one sale per customer!?"*

   Your primary key field needs work. You may need a compound primary key—CUSTOMER NUMBER and some other field or fields. In this case, CUSTOMER NUMBER, PRODUCT NUMBER, and DATE OF SALE might provide a unique combination of values—or you might consider using an INVOICE NUMBER field as a key.

3. *"My query says 'Enter Parameter Value' when I run it. What is that?"*

   This symptom, 99 times out of 100, indicates you have an expression in a Criteria or a Calculated Field, and *you misspelled a field name in the expression*. Access is very fussy about spelling. For example, Access is case sensitive. Furthermore, if you put a space in a field name when you define the table, then you must put a space in the field name when you reference it in a query expression. Fix the typo in the query expression.

   This symptom infrequently appears when you have a calculated field in a query, and you elect *not* to show the value of the calculated field in the query output. (You clicked off the Show box for the calculated field.) To get around this problem, click Show back on.

4. *"I'm getting a fantastic number of rows in my query output—many times more than I need. Most of the rows are duplicates!"*

   This symptom is usually caused by a failure to link together all tables you brought into the top half of the query generator. The solution is to use the manual click-and-drag method. Link the fields (usually primary key fields) with common *values* between tables. (Spelling of the field names is irrelevant because the link fields need not be spelled the same.)

5. *"For the most part, my query output is what I expected, but I am getting one or two duplicate rows."*

   You may have linked too many fields between tables. Usually only a single link is needed between two tables. It's unnecessary to link each common field in all combinations of tables; usually it's enough to link the primary keys. A layman's explanation for why over-linking causes problems is that excess linking causes Access to "overthink" the problem and repeat itself in its answer.

On the other hand, you might be using too many tables in the query design. For example, you brought in a table, linked it on a common field with some other table, but then did not use the table. You brought down none of its fields and/or you used none of its fields in query expressions. Therefore, get rid of the table, and the query should still work. Try doing this to see whether the few duplicate rows disappear: Click the unneeded table's header in the top of the QBE area and press the Delete key.

6. *"I expected six rows in my query output, but I only got five. What happened to the other one?"*

   Usually this indicates a data-entry error in your tables. When you link together the proper tables and fields to make the query, remember that the linking operation joins records from the tables *on common values* (*equal* values in the two tables). For example, if a primary key in one table has the value "123", the primary key or the linking field in the other table should be the same to allow linking. Note that the text string "123" is not the same as the text string "123 "—the space in the second string is considered a character too! Access does not see unequal values as an error: Access moves on to consider the rest of the records in the table for linking. Solution: Look at the values entered into the linked fields in each table and fix any data-entry errors.

7. *"I linked fields correctly in a query, but I'm getting the empty set in the output. All I get are the field name headings!"*

   You probably have zero common (equal) values in the linked fields. For example, suppose you are linking on Part Number (which you declared as text): In one field you have part numbers "001", "002", and "003", and in the other table part numbers "0001", "0002", and "0003". Your tables have no common values, which means no records are selected for output. You'll have to change the values in one of the tables.

8. *"I'm trying to count the number of today's sales orders. A Sigma query is called for. Sales are denoted by an invoice number, and I made this a text field in the table design. However, when I ask the Sigma query to 'Sum' the number of invoice numbers, Access tells me I cannot add them up! What is the problem?"*

   Text variables are words! You cannot add words, but you can count them. Use the Count Sigma operator (not the Sum operator): Count the number of sales, each being denoted by an invoice number.

9. *"I'm doing Time arithmetic in a calculated field expression. I subtracted the Time In from the Time Out and I got a decimal number! I expected 8 hours, and I got the number .33333. Why?"*

   [Time Out] – [Time In] yields the decimal percentage of a 24-hour day. In your case, 8 hours is one-third of a day. You must complete the expression by multiplying by 24: ([Time Out] – [Time In]) * 24. Don't forget the parentheses!

10. *"I formatted a calculated field for currency in the query generator, and the values did show as currency in the query output; however, the report based on the query output does not show the dollar sign in its output. What happened?"*

    Go into the report Design View. There is a box in one of the panels representing the calculated field's value. Click the box and drag to widen it. That should give Access enough room to show the dollar sign, as well as the number, in output.

11. *"I told the Report Wizard to fit all my output to one page. It does print to just one page. But some of the data is missing! What happened?"*

Access fits the output all on one page by *leaving data out*! If you can stand to see the output on more than one page, click off the "Fit to a Page" option in the Wizard. One way to tighten output is to go into the Design View and remove space from each of the boxes representing output values and labels. Access usually provides more space than needed.

12. *"I grouped on three fields in the Report Wizard, and the Wizard prints the output in a staircase fashion. I want the grouping fields to be on one line! How can I do that?"*

    Make adjustments in the Design View. See the Reports section of this tutorial for instruction.

13. *"When I create an Update query, Access tells me that zero rows are updating, or more rows are updating than I want. What is wrong?"*

    If your Update query is not correctly set up, for example, if the tables are not joined properly, it will either try not to update anything, or it will update all the records. Check the query, make corrections, and run it again.

14. *"After making a Summation Query with a Sum in the Group By row and saving that query, when I go back to it, the Sum field now says Expression, and Sum is put in the field name box. Is this wrong?"*

    Access sometimes changes that particular statistic when the query is saved. The data remains the same, and you can be assured your query is correct.

Preliminary Case: The Fitness Club

SETTING UP A RELATIONAL DATABASE TO CREATE TABLES, A FORM, QUERIES, AND A REPORT

PREVIEW

In this case, you'll create a relational database for exercise classes at a fitness club. First, you'll create four tables and populate them with data. Then you'll create a form for easy class registration and three queries: a Parameter query, a Count query, and a query with a Calculated Field. Finally, you'll create a payroll report showing weekly pay for exercise instructors.

PREPARATION

- Before attempting this case, you should have some experience in using Microsoft Access.
- Complete any part of Access Tutorial B that your instructor assigns, or refer to the tutorial as necessary.

BACKGROUND

Your mom exercises at a local fitness club and complains that the fitness club doesn't do a good job of keeping track of exercise classes. Because you are home for summer vacation, you volunteer to create an Access database for the club to help them keep track of the exercise classes, students, and instructors. With your experience in Microsoft Access, you feel confident you can help the club. In return, the club owner will allow you to take free classes during the summer.

It is interesting that, the son of the fitness club's owner started this project last summer, but he didn't have time to finish it. He has already set up the Microsoft Access tables for you. There are four tables in the database:

1. The STUDENTS table keeps track of each student's ID number, full name, address, and telephone number.
2. The INSTRUCTORS table keeps track of all the instructors employed by the fitness club. The table lists each instructor's ID number, full name, complete address, and telephone number.
3. The CLASSES table describes all the fitness classes offered. For each class, the table lists a unique ID number, class name, class instructor, start and end times, day of the week on which the class is held, and the start date.
4. The REGISTRATION table records the registration of each class member by listing the student ID number and the class ID number.

The club owner has a few requirements of the database. First, he would like student registration to be easy. This task can be accomplished with a form. In addition, he would like the database to provide certain information so he can quickly give answers to students who inquire about classes. He requests the following:

- He would like to be able to type an instructor's name and see the class or classes that the instructor is currently teaching. This is in response to many clients who telephone and ask for classes taught by specific instructors.
- He would like to be able to figure out how many students are enrolled in a particular class so he knows whether a class is full.
- He would also like an easy way to determine the duration of a given class because prospective students always ask, "How long does it last?"

The owner would also like a weekly payroll report. Instructors are paid by the week, at an hourly rate, and the owner would like a weekly report of the classes each instructor teaches, their pay for each class, and their total pay for the week.

ASSIGNMENT 1 CREATING TABLES

Use Microsoft Access to create the four tables with the fields shown in Figures 1–1 through 1–4 and discussed in the Background section. Populate the database tables with data as shown. Add your name to the STUDENTS table as Student ID W-9, complete with your address and telephone number.

Student ID	Student Name	Street Address	City	State	Zip	Phone
B-10	Joe Breskley	5 Fifth Ave	Newark	DE	19711	(302)738-1220
F-5	Mary Beth Fairnham	1 Main St	Newark	DE	19713	(302)454-0009
H-10	George Hammer	6 Stewart St	Elsmere	DE	19832	(302)998-7682
J-11	Lou Jones	10 Lindon Rd	Wilmington	DE	19808	(302)454-3399
J-6	Petra Jordan	89 Oak Ave	Newark	DE	19711	(302)731-9821
L-16	David Luber	90 Harbor Ct	Wilmington	DE	19808	(302)998-3562
S-59	Ann Sweeney	6 Elm St	Wilmington	DE	19808	(302)998-1114
S-62	Pat Short	515 Oak Ave	Newark	DE	19713	(302)733-8732
W-9	Your Name					

Figure 1-1 The STUDENTS table

Instructor ID	Instructor First Name	Instructor Last Name	Street Address	City	State	Zip	Phone
12	Craig	Peterson	6 Pike Lane	Newark	DE	19711	(302)738-0011
15	Susan	Braun	1001-A Littlewood Ct	Wilmington	DE	19808	(302)998-4556
18	Paul	Sandia	15 Maple Pl	Hockessin	DE	19801	(302)856-9899
20	Betsy	Zeeba	25 West Ave	Newark	DE	19713	(302)731-4380

Figure 1-2 The INSTRUCTORS table

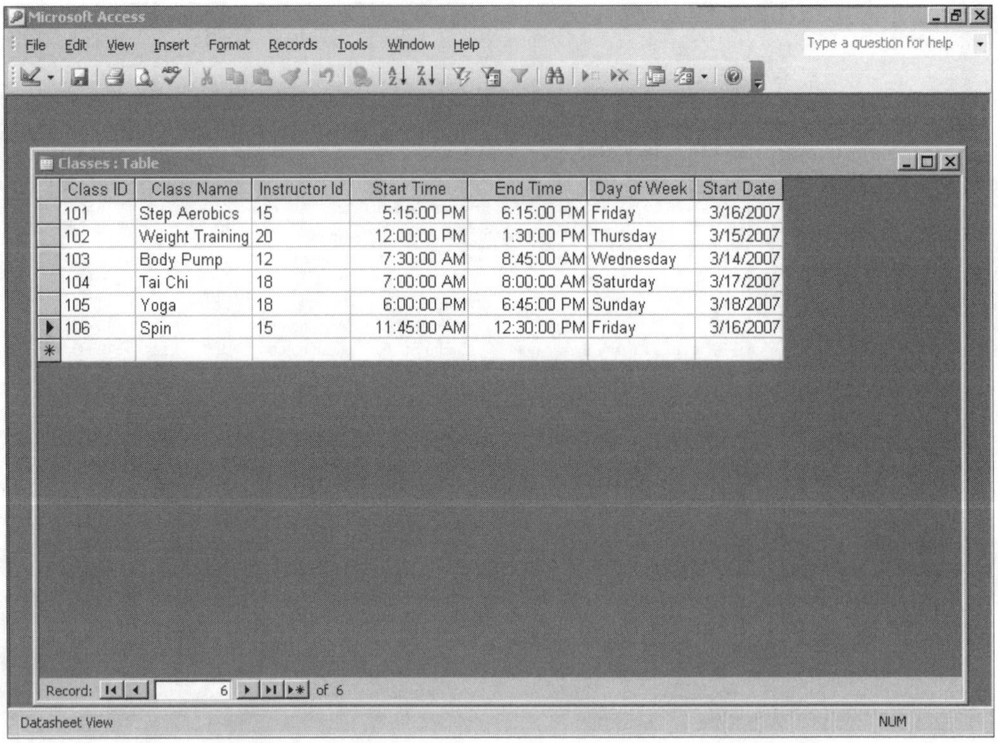

Class ID	Class Name	Instructor Id	Start Time	End Time	Day of Week	Start Date
101	Step Aerobics	15	5:15:00 PM	6:15:00 PM	Friday	3/16/2007
102	Weight Training	20	12:00:00 PM	1:30:00 PM	Thursday	3/15/2007
103	Body Pump	12	7:30:00 AM	8:45:00 AM	Wednesday	3/14/2007
104	Tai Chi	18	7:00:00 AM	8:00:00 AM	Saturday	3/17/2007
105	Yoga	18	6:00:00 PM	6:45:00 PM	Sunday	3/18/2007
106	Spin	15	11:45:00 AM	12:30:00 PM	Friday	3/16/2007

Figure 1-3 The CLASSES table

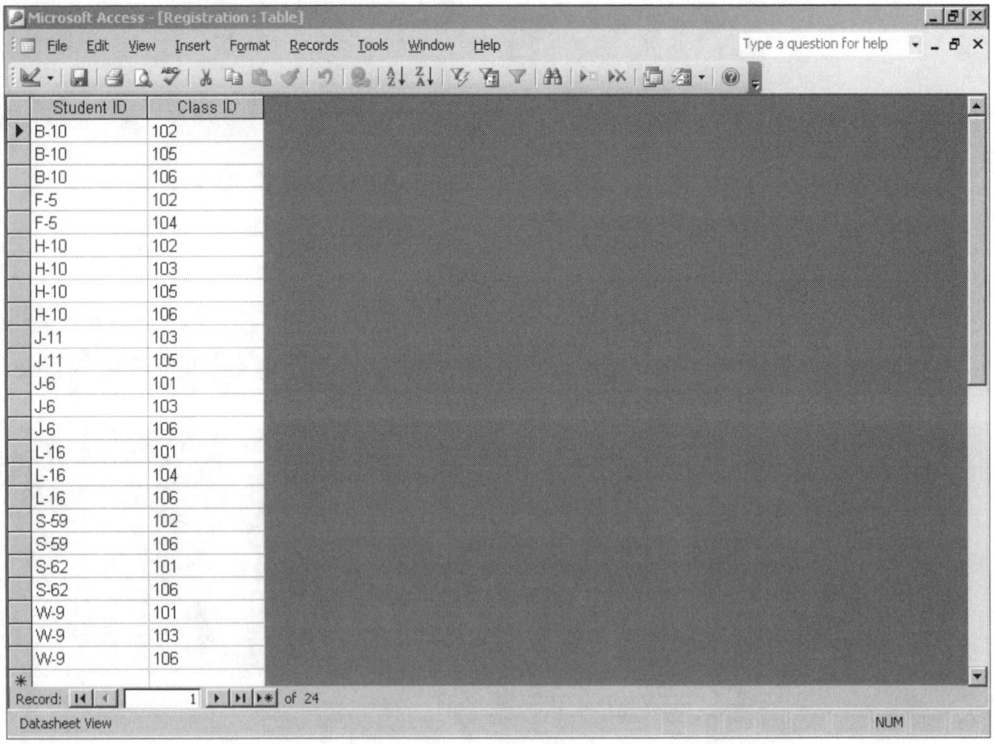

Student ID	Class ID
B-10	102
B-10	105
B-10	106
F-5	102
F-5	104
H-10	102
H-10	103
H-10	105
H-10	106
J-11	103
J-11	105
J-6	101
J-6	103
J-6	106
L-16	101
L-16	104
L-16	106
S-59	102
S-59	106
S-62	101
S-62	106
W-9	101
W-9	103
W-9	106

Figure 1-4 The REGISTRATION table

ASSIGNMENT 2 CREATING A FORM, QUERIES, AND A REPORT

Assignment 2A: Creating a Form

Create a form so the owner can easily input registration data. Using only the REGISTRATION table and the Form Wizard, choose Column layout and any Style. Save the form as Registration. View one record and print only one record.

Assignment 2B: Creating a Parameter Query

Create a Parameter Query that prompts for an Instructor's first name. Most students know instructors by their first name only and when telephoning the club, they request that instructor by first name. Create a query that allows input of that name. The output should display the Instructor First Name, Class Name, Day of Week, Start Date, and Start Time. Save the query as Which Class.

If a student asks which classes "Craig" is teaching, the output answer would resemble that shown in Figure 1-5.

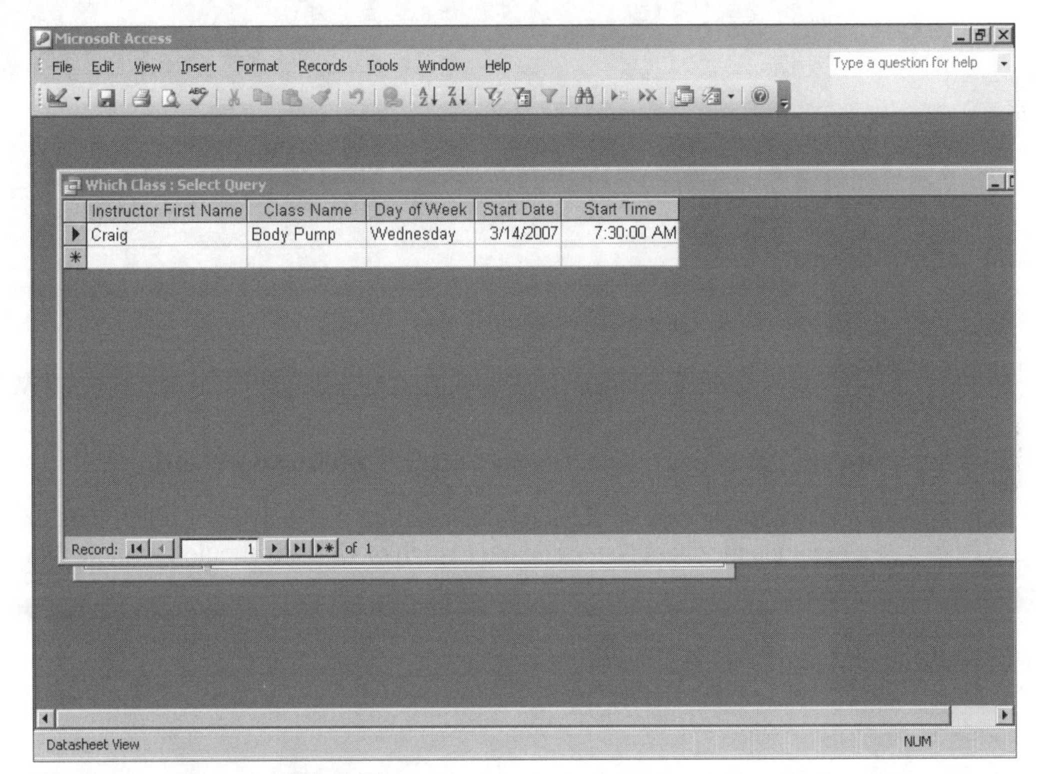

Figure 1-5 Query 1: Which Class

Run the Parameter Query with the input of "Craig". Print the results.

Assignment 2C: Creating a Count Query

Create a query that counts the number of students registered for the Weight Training class. Your output should resemble that shown in Figure 1-6. Print the output.

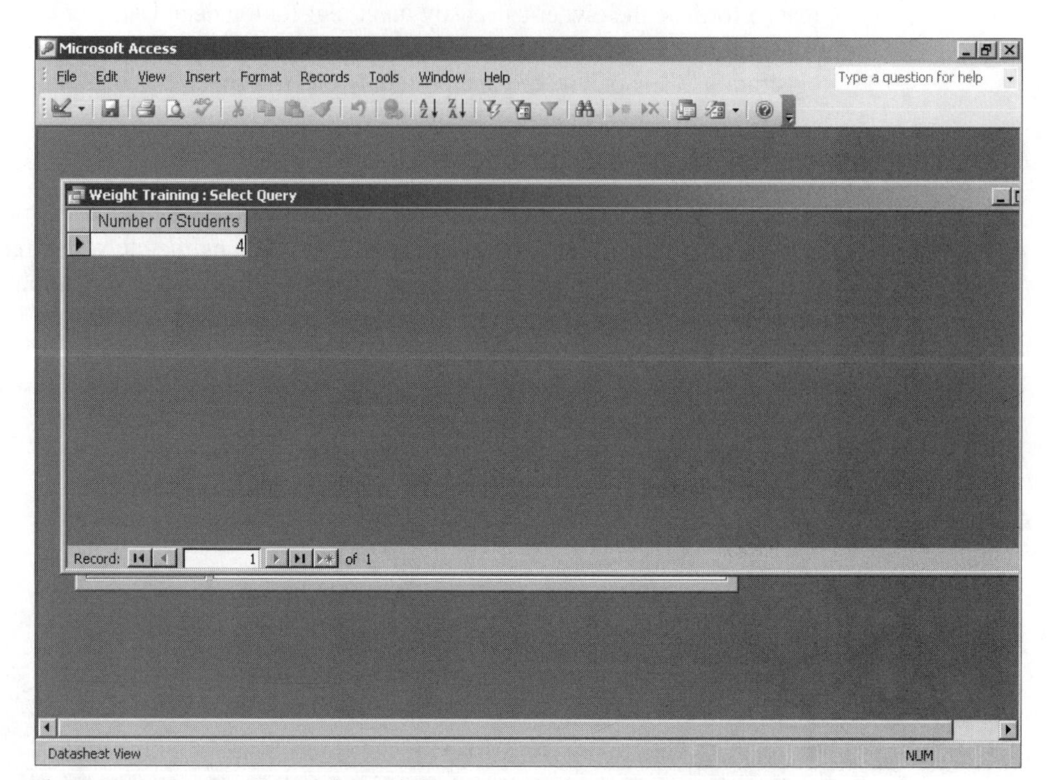

Figure 1-6 Query 2: Weight Training

Note the heading change (Number of Students). Save the query as Weight Training.

Assignment 2D: Creating a Query with a Calculated Field

Create a query that calculates the length of the Body Pump class in hours. Your output should include the Class Name and the Length of Class; it should resemble that shown in Figure 1-7. Print the output.

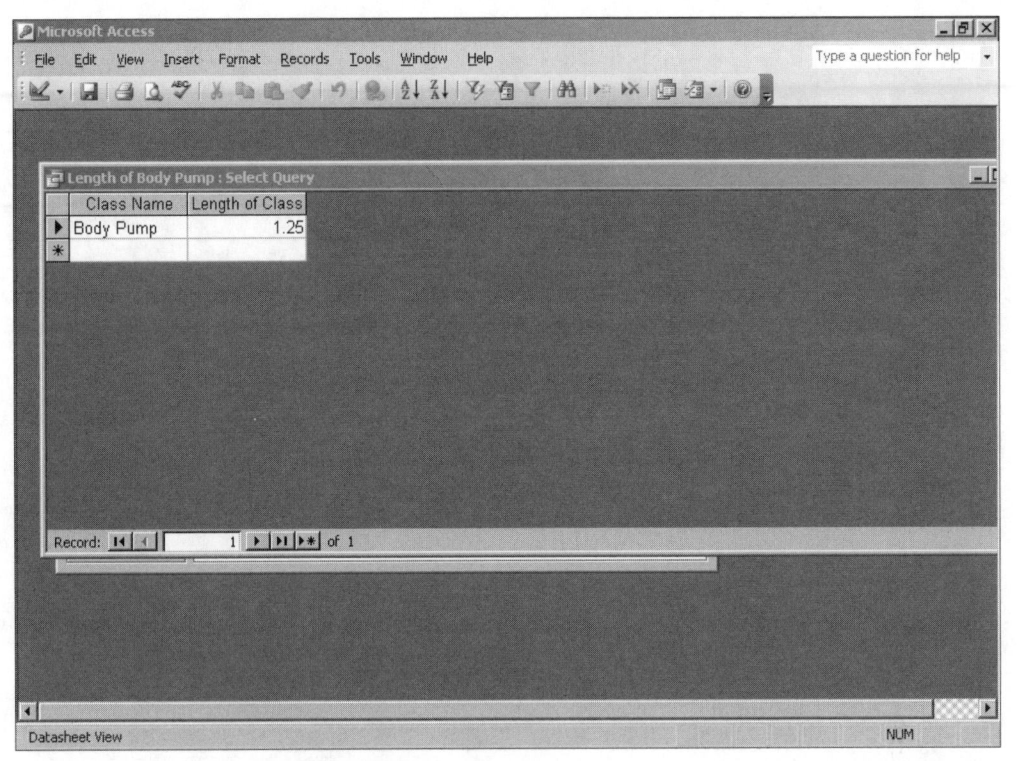

Figure 1-7 Query 3: Length of Body Pump

Note the heading of the calculated field. Save the query as Length of Body Pump.

Assignment 2E: Generating a Weekly Payroll Report

Generate a report that shows the weekly pay for each instructor for each class they teach. To create the report, you need to do the following:

- You must first create a query that brings all the data together for the report. You will need these headings: Instructor Name, Instructor Address, Class Name, and Pay. Your report should also show Total Weekly Pay for each instructor. The Pay column is a calculated field. Assume that the instructors earn $25 per hour.
- Using the Report Wizard, base your report on the query you made in the previous bulleted item.
- Group on Instructor First Name. Give the report a title of Weekly Instructor Pay.
- In the Design View, move the remaining instructor fields (last name, all address fields) to the Instructor First Name Header so information is not repeated if the instructor taught more than one class.
- Adjust the page header and any other parts of the report to resemble that shown in Figure 1-8. (Only the top portion of the report is showing in the figure.)
- Use Print Preview before printing to make sure the report looks correct.

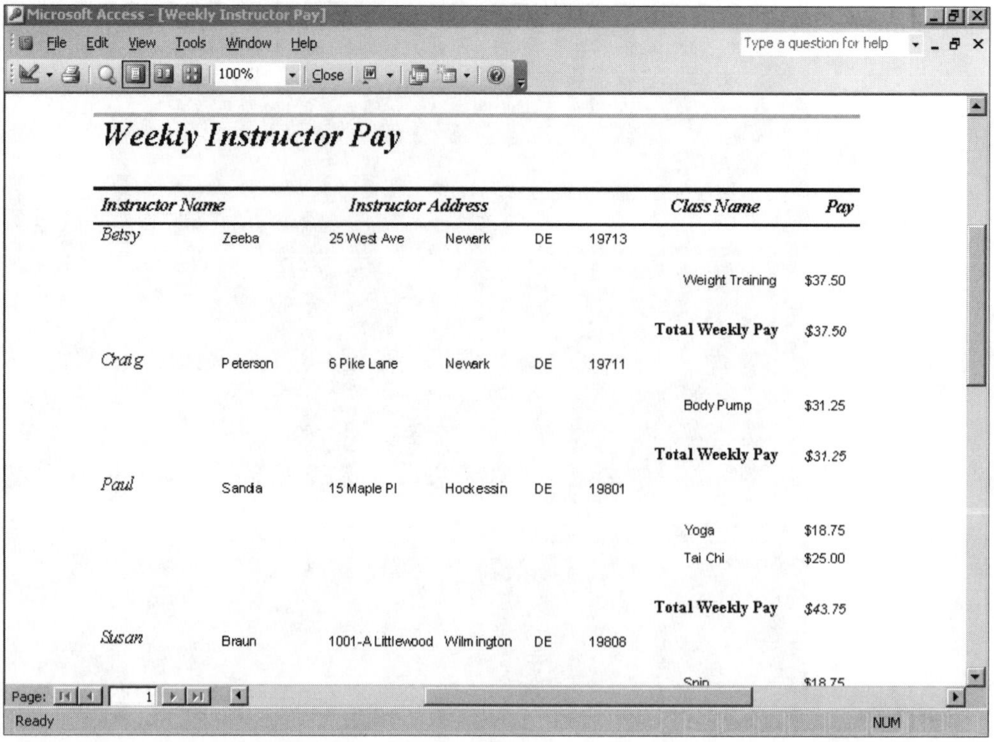

Microsoft Access - [Weekly Instructor Pay]

File Edit View Tools Window Help

Weekly Instructor Pay

Instructor Name		Instructor Address				Class Name	Pay
Betsy	Zeeba	25 West Ave	Newark	DE	19713		
						Weight Training	$37.50
						Total Weekly Pay	*$37.50*
Craig	Peterson	6 Pike Lane	Newark	DE	19711		
						Body Pump	$31.25
						Total Weekly Pay	*$31.25*
Paul	Sanda	15 Maple Pl	Hockessin	DE	19801		
						Yoga	$18.75
						Tai Chi	$25.00
						Total Weekly Pay	*$43.75*
Susan	Braun	1001-A Littlewood	Wilmington	DE	19808		
						Spin	$18.75

Page: 1

Ready NUM

Figure 1-8 Report: Weekly Instructor Pay

If you are working with a disk or CD, make sure you close the database file before removing your disk or CD.

DELIVERABLES

Assemble the following deliverables for your instructor:

1. Printouts of four tables
2. Form: Registration
3. Query 1: Which Class
4. Query 2: Weight Training
5. Query 3: Length of Body Pump
6. Report: Weekly Instructor Pay
7. Any other required tutorial printouts, tutorial disk, or CD

Staple all pages together. Put your name and class number at the top of the page. If required, make sure your disk or CD is labeled.

The Flying Club

2
CASE

Designing a Relational Database to Create Tables, Forms, Queries, and a Report

Preview

In this case, you'll design a relational database for a flying club. After your database design is completed and correct, you will create database tables and populate them with data. Then you will produce two forms, three queries, and one report. Members will use one form to reserve aircraft; mechanics will use the other form to log maintenance and repairs. The queries will answer the following questions: Who are the members in Pennsylvania? How many times has each aircraft group been rented? What is the bill for a particular member? The report will list the flying hours logged for each member.

Preparation

- Before attempting this case, you should have some experience in database design and in using Microsoft Access.
- Complete any part of Database Design Tutorial A that your instructor assigns.
- Complete any part of Access Tutorial B that your instructor assigns, or refer to the tutorial as necessary.
- Refer to Tutorial E as necessary.

BACKGROUND

Years ago, your uncle flew military aircraft, and now that he's retired, he wants to indulge his love of flying. He recently joined a local flying club that owns several small aircraft that they maintain, service, and rent to club members—who are all licensed pilots. For club members, renting an airplane by the hour is much more economical than buying and maintaining a private aircraft. Some members also offer flying lessons on the side.

You're visiting your uncle at his New Jersey home for the summer, and the two of you have gone on several flying trips. As a result, you've also become interested in flying. When talking to the manager of the flying club, Sandy Black, you've expressed an interest in taking flying lessons but claim you are an impoverished college student. She inquires about your major and the courses you are taking at university. She finds out that you are proficient in database design and Microsoft Access, so she offers you a deal: Create a database for the club and she will give you five free flying lessons. You agree and begin the project.

Sandy Black wants the database to keep track of current club members—each member's ID number, address, and telephone number. It also is important to note whether a member has an up-to-date pilot's license. A pilot's license does not expire, but to keep it current, the pilot must log a certain amount of flying time. In addition, Sandy wants to use the database to perform specific tasks.

First, Sandy wants making aircraft reservations to be easy. A reservations form for members can accomplish that task. Eventually, the reservation form will be put on the Internet so members can access it from home.

Second, Sandy wants mechanics who service the aircraft to use a form to record service and repairs.

Third, Sandy would like to use the database to help recruit new members. With a slight economic downturn, the club has lost a few members. The club would like to target potential members in nearby Pennsylvania, because studies show that area's population tends to have more disposable income. Sandy would like to see a list of all current members who live in Pennsylvania so she can call them and encourage them to recruit some of their pilot friends to join the club.

Fourth, Sandy would like to track aircraft rental. The club has four groups of airplanes—Cessna, Twin, Tailwheel, and Piper—with multiple airplanes in each group. Sandy would like to track how many times each group of aircraft is rented so that if new aircraft are to be purchased, she can buy an airplane in the group that best serves members' needs.

Fifth, Sandy wants to track fees. Some members use aircraft quite often but are delinquent in paying their fees. Sandy would like to be able to type in a member's last name and see the number of times the member has rented aircraft and the amount of money owed to the club for those rentals. Members pay rental fees every two months. It is easier to operate the club with this type of billing system so the club doesn't have to deal with frequent payments.

Sixth and finally, the club needs to keep track of all the flying hours logged by the members. This is not only to ensure that their pilot's license stays current, but also the club rules state that if a member logs three hours per month, he or she will get a safety incentive rebate on their annual membership dues. Sandy would like this information in a report format.

ASSIGNMENT 1 CREATING THE DATABASE DESIGN

In this assignment, you will design your database tables on paper, using a word-processing program. Pay close attention to the tables' logic and structure. Do not start your Access code (Assignment 2) before getting feedback from your instructor on Assignment 1. Keep in mind that you need to look at what is required in Assignment 2 to design your fields and tables properly. It's good programming practice to look at the required outputs before designing your database. When designing the database, observe the following guidelines:

- First, determine the tables you'll need by listing on paper the name of each table and the fields that it should contain. Avoid data redundancy. Do not create a field if it could be created by a "calculated field" in a query.
- Include a logical field that answers the question "Current License?"
- You'll need a transaction table. Avoid duplicating data.
- Document your tables by using the Table facility of your word processor. Your word-processed tables should resemble the format of the table in Figure 2-1.
- You must mark the appropriate key field(s). You can designate a key field by an asterisk (*) next to the field name. Keep in mind that some tables need a compound primary key to identify a record uniquely within a table.
- Print the database design.

TABLE NAME	
Field Name	***Data Type (text, numeric, currency, etc.)***
...	...
...	...

Figure 2-1 Table design

NOTE

Have your design approved before beginning Assignment 2; otherwise, you might need to redo Assignment 2.

Assignment 2 Creating the Database and Making Queries and a Report

In this assignment, you will first create database tables in Access and populate them with data. Then you will create two forms, three queries, and a report.

Assignment 2A: Creating Tables in Access

In this part of the assignment, you will create your tables in Access. Use the following guidelines:

- Type records into the tables, using the members' names and addresses shown in Figure 2-2. Add your name and address as an additional member. Assume all pilots' licenses are up-to-date. Make up member IDs and telephone numbers.
- Assume there are four different groups of aircraft: Twin, Piper, Cessna, and Tailwheel. For each airplane, make up the year in which it was manufactured, tail number (unique ID number), and model number. Assume the rental cost ranges from $65 to $246 per hour.

- Have each club member rent aircraft at least once during a two-month time period. Create data for those two months (at least 20 entries). Assume that the members are billed for their aircraft usage at the end of a two-month period.
- Appropriately limit the size of the text fields; for example, a telephone number does not need to be the default setting of 50 characters in length.
- Print all tables.

Last Name	First Name	Address	City	State	Zip
Burham	Luke	212 Wedgewood Rd	Kennet Square	PA	19348
Doorey	Leon	9 Sandalwood Dr	Avondale	PA	19311
Mattern	Luis	205 Hanover Pl	Avondale	PA	19311
Seals	Lauren	163 Darling Rd	Kennet Square	PA	19348
Webster	Sally	15 Anglin Dr	Wilmington	DE	19808
Ward	Harry	11 Chapel Rd	Wilmington	DE	19808
Seever	Patti	19 Danvers Circle	Kennet Square	PA	19348
Poplawski	Meredith	281 Beverly Rd	Elkton	MD	21119
Meartz	Maria	511 Sparrow Ct	Wilmington	DE	19808
Lee	Gregory	139 Boyer Way	Avondale	PA	19311

Figure 2-2 Data: member data

Assignment 2B: Creating Forms, Queries, and a Report

There are two forms, three queries, and one report to generate, as outlined in the background of this case.

Form 1: Registrations

Create a form based on the table that takes reservations. Include information such as reservation number, tail number, member ID, and the date and time of pick-up and drop-off. Save the form as Reservations. Print any one record from that form.

Form 2: Maintenance

Create a form to record maintenance records on the aircraft. Include in the form the tail number of the aircraft, the date of maintenance, and details of the maintenance. Save the form as Maintenance. Print any one record from the form.

Query 1: Members in PA

Create a query called Members in PA. The output of the query should have column heads for First Name, Last Name, Address, City, State, and Telephone number of only of those members who live in PA. Your output should resemble that shown in Figure 2-3.

Microsoft Access

File Edit View Insert Format Records Tools Window Help

Members in PA : Select Query

First Name	Last Name	Address	City	State	Telephone
Luke	Burham	212 Wedgewood Rd	Kennet Square	PA	(610)292-1779
Leon	Doorey	9 Sandelwood Dr	Avondale	PA	(610)366-1929
Luis	Mattern	205 Hanover Pl	Avondale	PA	(610)366-0159
Lauren	Seals	163 Darling Rd	Kennet Square	PA	(610)369-9550
Patti	Seever	19 Danvers Circle	Kennet Square	PA	(610)544-0923
Gregory	Lee	139 Boyer Way	Avondale	PA	(610)655-0936

Record: 6 of 6

Datasheet View NUM

Figure 2-3 Query 1: Members in PA

Case 2

Query 2: Rentals by Group

Create a query called Rentals by Group. Show each group of aircraft and the number of times it has been rented in a two-month period. Be careful to change the column headings so your output looks like that shown in Figure 2-4, although your data will be different.

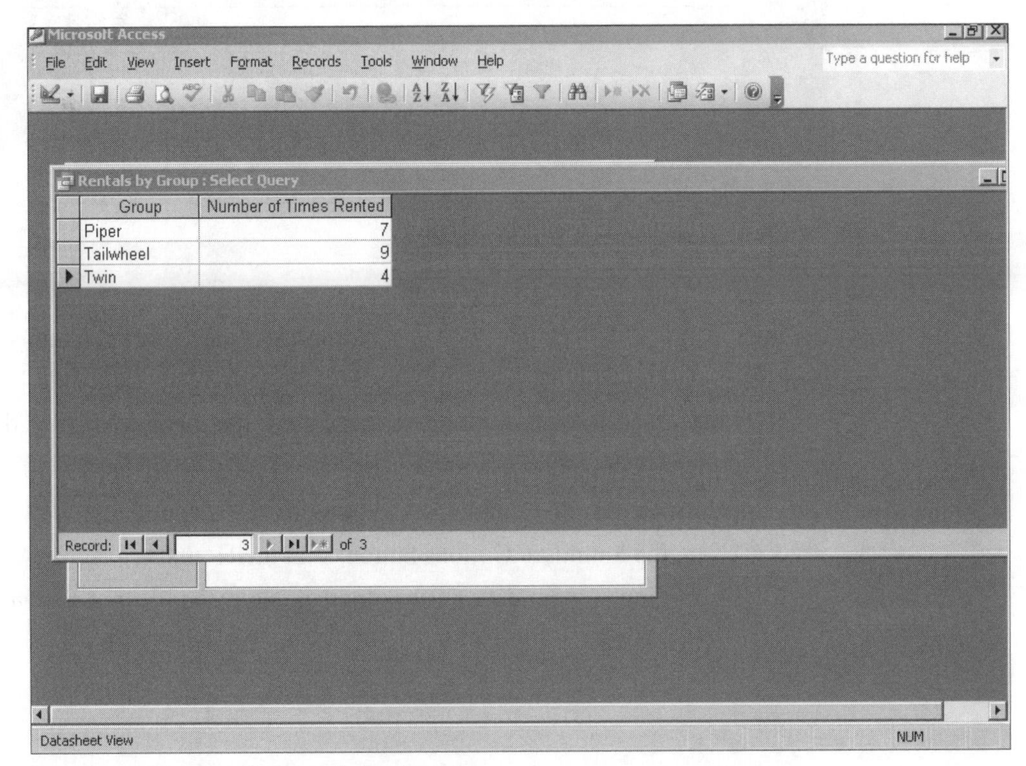

Group	Number of Times Rented
Piper	7
Tailwheel	9
Twin	4

Figure 2-4 Query 2: Rentals by Group

Query 3: Member Account

Create a query called Member Account:

- First, create a parameter query that will provide information about each member's rentals.
- The query should prompt the user for Member Last Name input.
- The output should show column headers for the member's Last Name, First Name, plane Tail #, plane Group, Date of rental, and Rental Fee per time.
- Your data might differ, but if you input the last name "Mattern," your output should resemble the format shown in Figure 2-5.

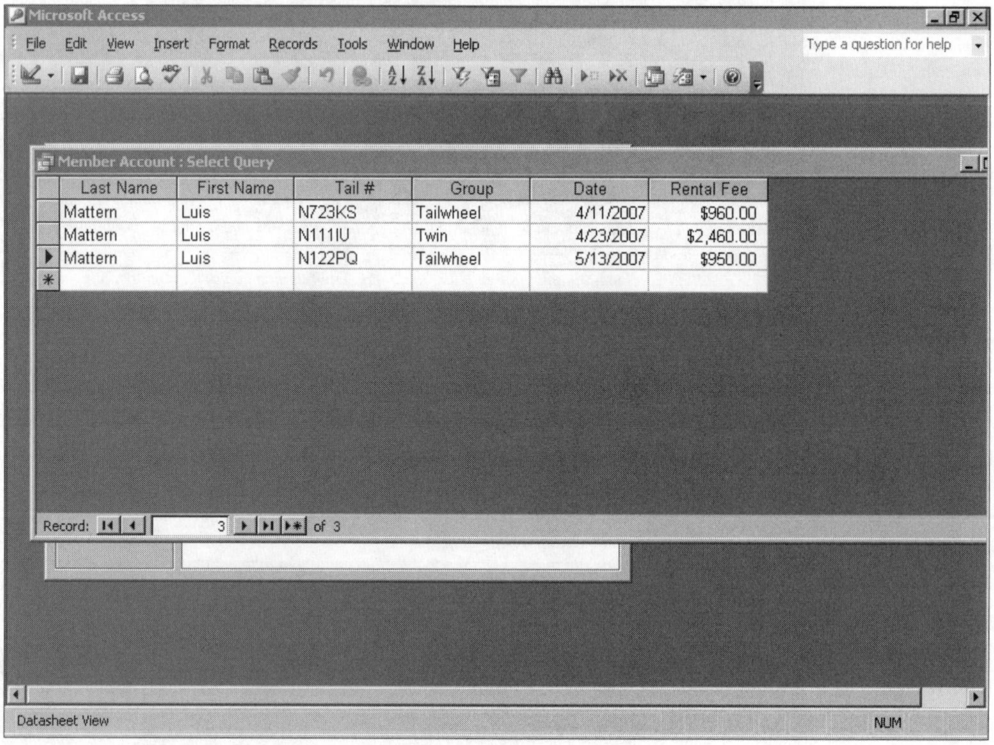

Last Name	First Name	Tail #	Group	Date	Rental Fee
Mattern	Luis	N723KS	Tailwheel	4/11/2007	$960.00
Mattern	Luis	N111IU	Twin	4/23/2007	$2,460.00
Mattern	Luis	N122PQ	Tailwheel	5/13/2007	$950.00

Figure 2-5 Query 3: Member Account

Report: Hours Logged by Member

Create a report named Hours Logged by Member. Your report's output should show headers for Last Name, First Name, Date, Group, and Hours. Use the following procedure:

1. First, create a query for input to that report, including a calculated field called Hours.
2. Group on the field Last Name.
3. Adjust the design of the report to bring the First Name into the Last Name header.
4. Using the Summary Options button, sum the Hours.
5. Depending on your data, your output should resemble that shown in Figure 2-6.

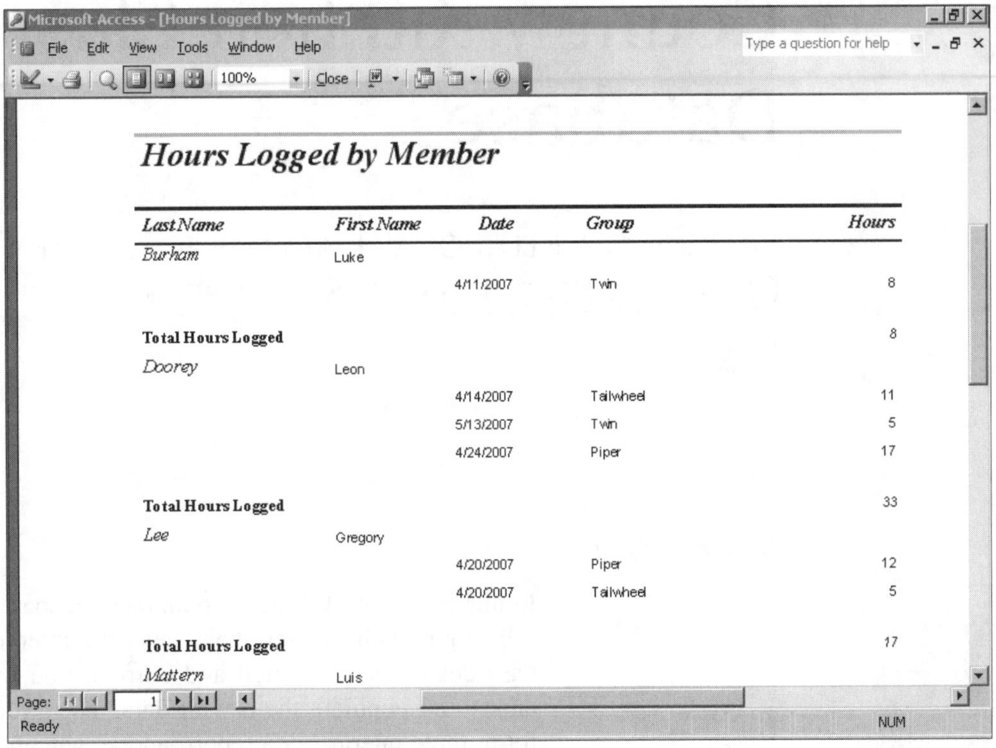

Microsoft Access - [Hours Logged by Member]

File Edit View Tools Window Help

Hours Logged by Member

LastName	First Name	Date	Group	Hours
Burham	Luke			
		4/11/2007	Twin	8
Total Hours Logged				8
Doorey	Leon			
		4/14/2007	Tailwheel	11
		5/13/2007	Twin	5
		4/24/2007	Piper	17
Total Hours Logged				33
Lee	Gregory			
		4/20/2007	Piper	12
		4/20/2007	Tailwheel	5
Total Hours Logged				17
Mattern	Luis			

Figure 2-6 Report: Hours Logged by Member

ASSIGNMENT 3 MAKING A PRESENTATION

Create a presentation that explains the database to your club members. Include the design of your database tables and how to use the database. Your presentation should take fewer than 10 minutes, including a brief question-and-answer period.

DELIVERABLES

1. Word-processed design of tables
2. Tables created in Access
3. Form 1: Registrations (one record printed)
4. Form 2: Maintenance
5. Query 1: Members in PA
6. Query 2: Rentals by Group
7. Query 3: Member Account (with one last name)
8. Report: Hours Logged by Member
9. Presentation materials
10. Any other required tutorial printouts, tutorial disk, or tutorial CD

Staple all pages together. Put your name and class number at the top of the page. Make sure your disk or CD is labeled, if required.

CASE

Country Kitchen Classics Database

Designing a Relational Database to Create Tables, a Form, Queries, a Report, and a Switchboard

Preview

In this case, you'll design a relational database for a business that sells high-quality frozen meals over the Internet. After your database design is completed and correct, you will create database tables and populate them with data. Then you will produce one form, three queries, one report, and a switchboard. The form will show the customer's order and billing information. The queries will answer the following questions: What is the description of this meal? Which meals had the most sales today? Which meals must ship via overnight delivery? The report will summarize today's orders per customer and show each order's total price, including the delivery charges. You will create a switchboard to manage the database easily.

Preparation

- Before attempting this case, you should have some experience in database design and in using Microsoft Access.
- Complete any part of Database Design Tutorial A that your instructor assigns.
- Complete any part of Access Tutorial B that your instructor assigns, or refer to the tutorial as necessary.
- Refer to Tutorial E as necessary.

BACKGROUND

Your cousin Alice and her business partner, Rachel, plan to sell quality, gourmet frozen meals over the Internet. Alice and Rachel have been studying the market and see a need for these meals, especially among the elderly and busy working families. Alice knows that you are proficient in database design and Microsoft Access. She hires you for the summer to help create this exciting business.

Alice envisions the business to include these parameters: customers, meals, orders, and shipping costs.

- *Customers* will be ordering their meals via the Internet, so the business needs to track their name, shipping address, telephone number, credit card details, and e-mail address.
- *Meals* will be categorized by their unique ID number. Each meal has a name, specific price, and description, and this information must also appear in the database.
- *Orders* details need to include the specific meal(s), quantities, customer, and whether the order requires overnight shipping.
- *Shipping costs* includes two flat-rate costs: one for regular shipping and one for overnight shipping.

Alice wants the database to accomplish certain key tasks. First, she wants you to create an Internet order form for ordering meals. You will create the prototype in Access, and sales-office personnel will test it. When the form meets Alice's approval, she will hire additional programmers to migrate the form to the Web. The form should show basic order headings, such as Order Number, Customer ID, Date, and a check box to indicate whether the customer needs overnight delivery. In a subform, the order information required to order meals should include headings for Order Number, Meal ID, and Quantity. The form and subform are shown in Figure 3-3.

Alice also wants you to create three queries:

1. Potential customers might telephone the office and ask for more details about certain menu items. So Alice would like the office staff to be able to create queries that will display the description and price of each meal.
2. Alice would also like an easy way to calculate the number of meals sold on a particular day. She would like to be able to input the date and see a list showing meals with highest to lowest sales. This information is to ensure adequate inventory in the future.
3. The shipping clerk needs to see which of today's orders require overnight delivery so he can meet the pick-up deadline.

Finally, Alice requires you to create a daily report that lists each customer's ID number, last name, their order, the dollar amount of the order, shipping cost, and total order, which includes both meals and shipping costs.

You would like to make a good impression on your cousin because she is employing you for the summer. Thus, you also propose to create a switchboard for this database, so the customer service reps can easily access the form and report.

ASSIGNMENT 1 CREATING THE DATABASE DESIGN

In this assignment, you will design your database tables on paper, using a word-processing program. Pay close attention to the tables' logic and structure. Do not start your Access code (Assignment 2) before getting feedback from your instructor on Assignment 1. Keep in mind that you will need to look at what is required in Assignment 2 to design your fields and tables properly. It's good programming practice to look at the required outputs before designing your database. When designing the database, observe the following guidelines:

- First, determine the tables you'll need by listing on paper the name of each table and the fields that it should contain. Avoid data redundancy. Do not create a field if it could be created by a "calculated field" in a query.
- Include a logical field that answers the question "Overnight Delivery?"
- You'll need at least one transaction table. Avoid duplicating data. It is recommended that you use two tables to record transactions, one being the line-item table.
- Document your tables by using the Table facility of your word processor. Your word-processed tables should resemble the format of the table in Figure 3-1.
- You must mark the appropriate key field(s). You can designate a key field by an asterisk (*) next to the field name. Keep in mind that some tables need a compound primary key to uniquely identify a record within a table.
- Print out the database design.

TABLE NAME	
Field Name	*Data Type (text, numeric, currency, etc.)*
...	...
...	...

Figure 3-1 Table design

Have your design approved before beginning Assignment 2; otherwise, you may need to redo Assignment 2.

ASSIGNMENT 2 CREATING THE DATABASE AND THE FORM, QUERIES, REPORT, AND SWITCHBOARD

In this assignment, you will first create database tables in Access and populate them with data. Then you will create a form and subform, three queries, a report, and a switchboard.

Assignment 2A: Creating Tables in Access

In this part of the assignment, you will create your tables in Access. Use the following guidelines:

- Type records into the tables, using the customers' data shown in Figure 3-2. Add your name and address as an additional customer. Make up customer IDs, e-mail addresses, and credit card information.
- Assume there are nine different types of main-dish meals, each having a unique meal ID, price, and description. Have meal prices range from $9.50 to $15.00. Include a well-written description of each dish. (We are ignoring other meal components, such as desserts, to expedite typing data.)
- Have each customer buy at least one meal, and have most customers buy more than one meal type. Create data for today only. You can set the Date field on the table that takes the orders to have a default of today's date. Check the Help menu for details about how to do this.
- Create two flat-fee shipping rates: $6 for standard ground shipping and $15 for overnight delivery.
- Appropriately limit the size of the text fields; for example, a zip code does not need to be the default setting of 50 characters in length.
- Print all tables.

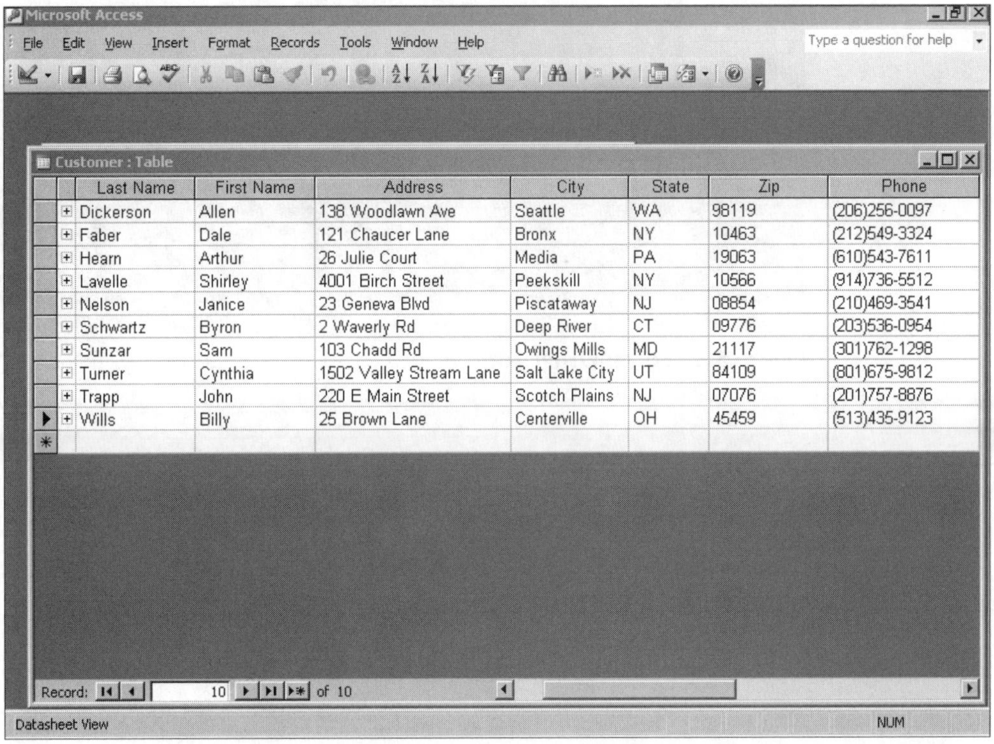

Microsoft Access

Customer : Table

Last Name	First Name	Address	City	State	Zip	Phone
Dickerson	Allen	138 Woodlawn Ave	Seattle	WA	98119	(206)256-0097
Faber	Dale	121 Chaucer Lane	Bronx	NY	10463	(212)549-3324
Hearn	Arthur	26 Julie Court	Media	PA	19063	(610)543-7611
Lavelle	Shirley	4001 Birch Street	Peekskill	NY	10566	(914)736-5512
Nelson	Janice	23 Geneva Blvd	Piscataway	NJ	08854	(210)469-3541
Schwartz	Byron	2 Waverly Rd	Deep River	CT	09776	(203)536-0954
Sunzar	Sam	103 Chadd Rd	Owings Mills	MD	21117	(301)762-1298
Turner	Cynthia	1502 Valley Stream Lane	Salt Lake City	UT	84109	(801)675-9812
Trapp	John	220 E Main Street	Scotch Plains	NJ	07076	(201)757-8876
Wills	Billy	25 Brown Lane	Centerville	OH	45459	(513)435-9123

Record: 10 of 10

Datasheet View

Figure 3-2 Data: Records for the CUSTOMER Table (does not include Customer ID, e-mail address, or credit card data)

Assignment 2B: Creating a Form and Subform, Queries, a Report, and a Switchboard

You will need to generate a form and subform, three queries, one report, and one switchboard, as outlined in the Background section of this case.

Form: Order Form

Create an order form and subform based on the process of taking orders. Call it Order Form. Print any one record from that form. The format of your form and subform should resemble that shown in Figure 3-3, but your data will be different.

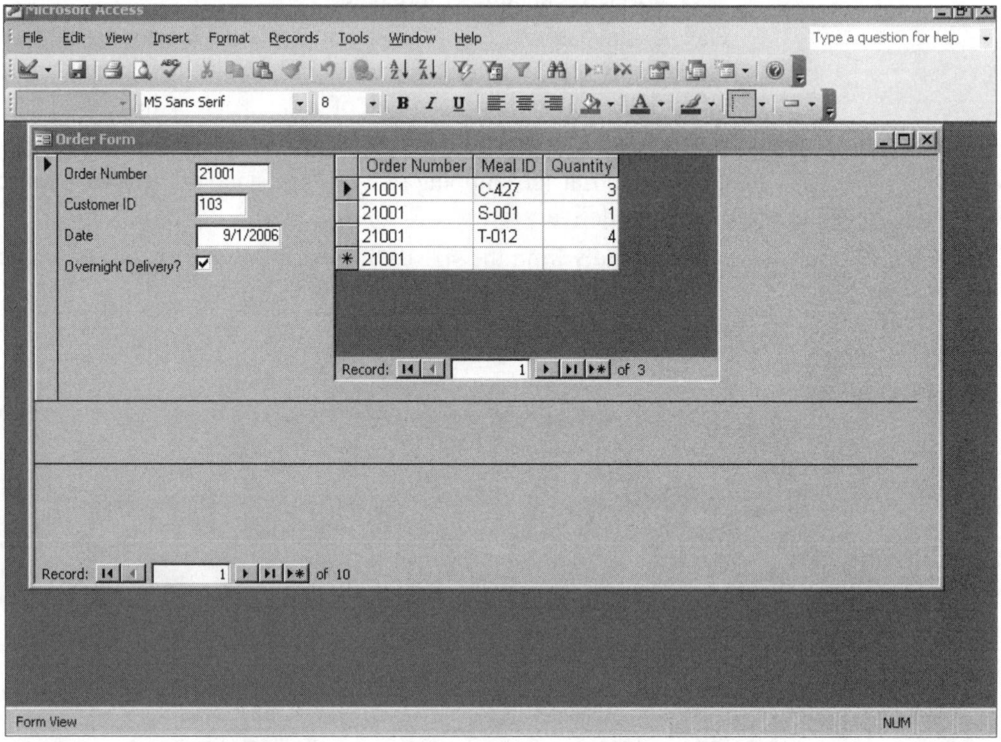

Microsoft Access

File Edit View Insert Format Records Tools Window Help

Order Form

Order Number 21001
Customer ID 103
Date 9/1/2006
Overnight Delivery? ☑

Order Number	Meal ID	Quantity
21001	C-427	3
21001	S-001	1
21001	T-012	4
21001		0

Record: 1 of 3

Record: 1 of 10

Form View

Figure 3-3 Form: Order Form

Query 1: Description of ______

Create a query called Description of BBQ Ribs (or "Description of ______ " any other dish you choose). The output of the query should show column headers for only the Meal ID, Meal Type, Price, and Description of the BBQ Ribs (or whatever dish you choose). Print that output. Your output format should resemble that shown in Figure 3-4.

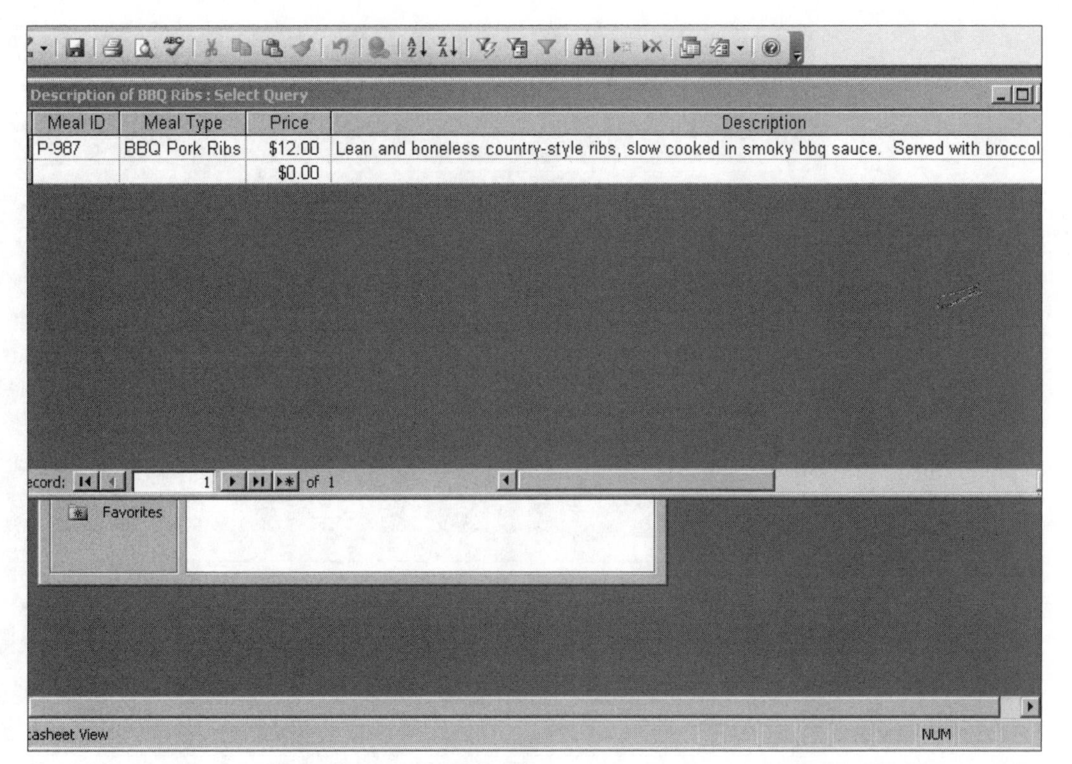

Figure 3-4 Query 1: Description of BBQ Ribs

Case 3

Query 2: Today's Meal Popularity

Create a query called Today's Meal Popularity. Column headers should show today's Date, Meal Type, and the Number of Meals sold today, listed from the high seller to the low seller. Be careful to change the column headings so your output format looks like that shown in Figure 3-5, although your data will be different. Print the output.

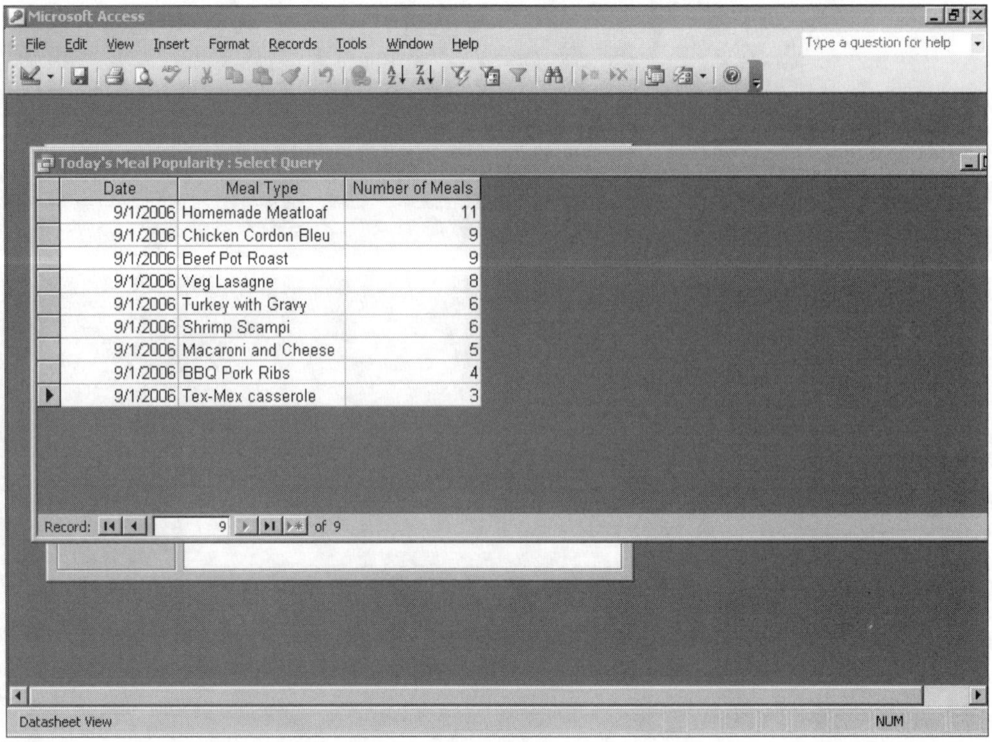

Date	Meal Type	Number of Meals
9/1/2006	Homemade Meatloaf	11
9/1/2006	Chicken Cordon Bleu	9
9/1/2006	Beef Pot Roast	9
9/1/2006	Veg Lasagne	8
9/1/2006	Turkey with Gravy	6
9/1/2006	Shrimp Scampi	6
9/1/2006	Macaroni and Cheese	5
9/1/2006	BBQ Pork Ribs	4
9/1/2006	Tex-Mex casserole	3

Figure 3-5 Query 2: Today's Meal Popularity

Query 3: Today's Overnight Deliveries

Create a query called Today's Overnight Deliveries. The query output should display the following column headers: Date, Customer ID, Last Name, First Name, Address, City, State, Zip, Phone, E-mail Address, Credit Card Details, Meal ID, and Quantity.

A portion of the output is shown in Figure 3-6. Your data will be different. Print your output.

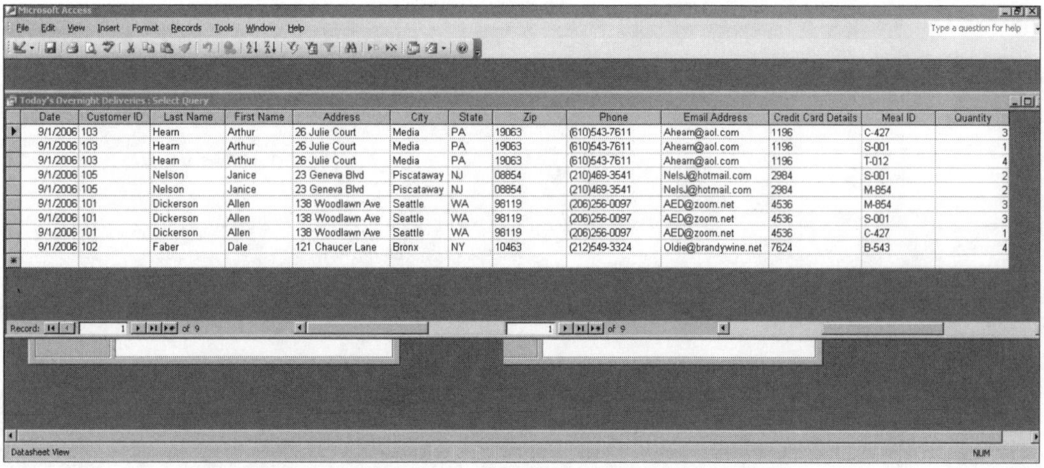

Date	Customer ID	Last Name	First Name	Address	City	State	Zip	Phone	Email Address	Credit Card Details	Meal ID	Quantity
9/1/2006	103	Hearn	Arthur	26 Julie Court	Media	PA	19063	(610)543-7611	Ahearn@aol.com	1196	C-427	3
9/1/2006	103	Hearn	Arthur	26 Julie Court	Media	PA	19063	(610)543-7611	Ahearn@aol.com	1196	S-001	1
9/1/2006	103	Hearn	Arthur	26 Julie Court	Media	PA	19063	(610)543-7611	Ahearn@aol.com	1196	T-012	4
9/1/2006	105	Nelson	Janice	23 Geneva Blvd	Piscataway	NJ	08854	(210)469-3541	NelsJ@hotmail.com	2984	S-001	2
9/1/2006	105	Nelson	Janice	23 Geneva Blvd	Piscataway	NJ	08854	(210)469-3541	NelsJ@hotmail.com	2984	M-854	2
9/1/2006	101	Dickerson	Allen	138 Woodlawn Ave	Seattle	WA	98119	(206)256-0097	AED@zoom.net	4536	M-854	3
9/1/2006	101	Dickerson	Allen	138 Woodlawn Ave	Seattle	WA	98119	(206)256-0097	AED@zoom.net	4536	S-001	3
9/1/2006	101	Dickerson	Allen	138 Woodlawn Ave	Seattle	WA	98119	(206)256-0097	AED@zoom.net	4536	C-427	1
9/1/2006	102	Faber	Dale	121 Chaucer Lane	Bronx	NY	10463	(212)549-3324	Oldie@brandywine.net	7624	B-543	4

Figure 3-6 Query 3: Today's Overnight Deliveries

Report: Summary of Today's Orders

Create a report Summary of Today's Orders. Use the following procedure:

1. First create a query as input to the report that includes Customer ID, Delivery Charge (which will be either regular or extra for the overnight shipping), Meal Type, Price, Quantity, and Total, which is a calculated field. Make sure the data is only for today.
2. In the report, sum the Total price of each customer's order, grouped by Customer ID, and call it Total Price of Order.
3. Title the report Summary of Today's Orders.
4. Adjust the Design View so your report looks professional and resembles the portion of the report shown in Figure 3-7. Print the report. Your data will differ.

Microsoft Access - [for report]

File Edit View Tools Window Help

Summary of Today's Orders

Customer ID	Delivery Charge	Meal Type	Price	Quantity	Total
101					
	$15				
		Chicken Cordon Bleu	$12.75	1	$12.75
		Shrimp Scampi	$15.00	3	$45.00
		Macaroni and Cheese	$10.00	3	$30.00
		Total Price of Order			$87.75
102					
	$15				
		Beef Pot Roast	$15.00	4	$60.00
		Total Price of Order			$60.00
103					
	$15				
		Shrimp Scampi	$15.00	1	$15.00

Page: 1

Ready NUM

Figure 3-7 Report: Summary of Today's Orders

Switchboard: Country Kitchen Classics

Create a Switchboard for easy access to the Order Form and Summary of Today's Orders report. Your Switchboard should resemble that shown in Figure 3-8.

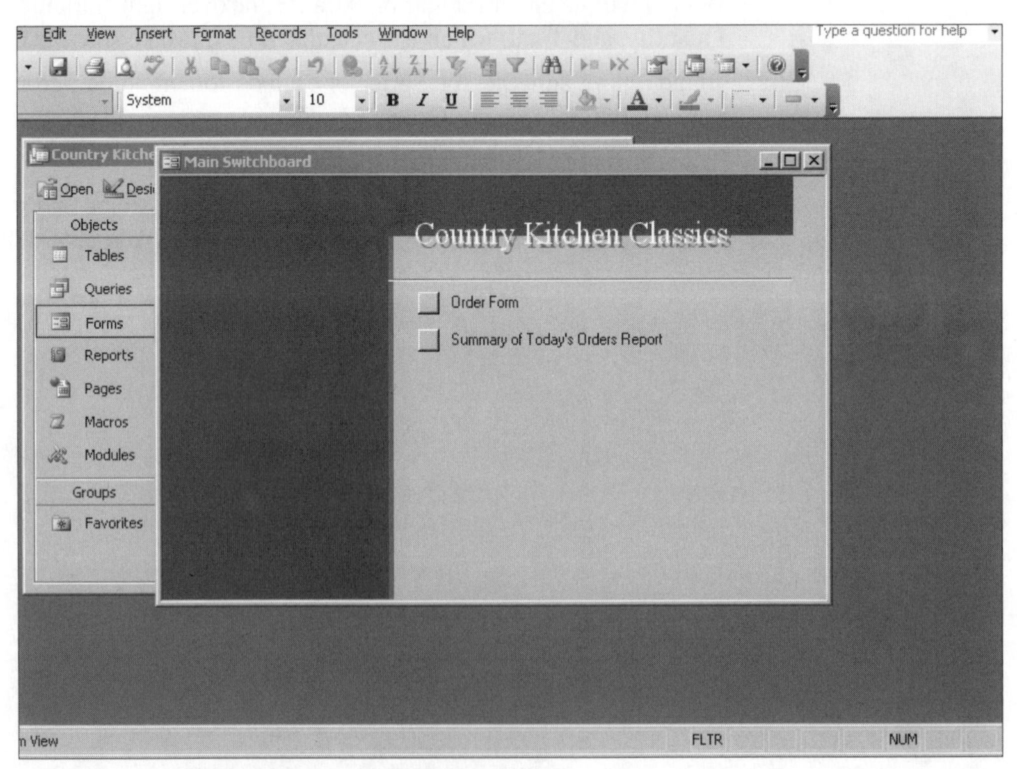

Figure 3-8 Switchboard: Country Kitchen Classics

ASSIGNMENT 3 MAKING A PRESENTATION

Create a presentation that explains the database to Alice and Rachel. Make sure you discuss the design of your database and how to use the forms, switchboard, and queries. Suggest future embellishments to the database that you might create next summer. Your presentation should take fewer than 10 minutes, including a brief question-and-answer period.

DELIVERABLES

1. Word-processed design of tables
2. Tables created in Access
3. Form: Order Form (1 record printed)
4. Query 1: Description of ______
5. Query 2: Today's Meal Popularity
6. Query 3: Today's Overnight Deliveries
7. Report: Summary of Today's Orders
8. Presentation materials
9. Any other required tutorial printouts or tutorial CD or disk, if required

Staple all pages together. Put your name and class number at the top of the page. Make sure your disk or CD is labeled, if required.

CASE

The Modeling Agency Database

Designing a Relational Database to Create Tables, Forms, Queries, a Report, and a Switchboard

Preview

In this case, you'll design a relational database for an agency that manages models. After your database design is completed and correct, you will create database tables and populate them with data. Then you will produce two forms, five queries, one report, and a switchboard. One form will record information about each new model and will including a photograph. The second form, which will contain a subform, will record customers' bookings for models. The queries will address a number of issues for the agency: What are the most frequently requested types of models, which models are the most frequently requested, which models fit a certain type, and which customers reimburse models for travel expenses? A fifth query will update models' fees. The report will summarize models' jobs and show the agency's fee for each job. Finally, a switchboard will allow easy access to the forms and report.

Preparation

- Before attempting this case, you should have some experience in database design and in using Microsoft Access.
- Complete any part of Database Design Tutorial A that your instructor assigns.
- Complete any part of Access Tutorial B that your instructor assigns, or refer to the tutorial as necessary.
- Refer to Tutorial E as necessary.

Background

Your Aunt Natasha runs a modeling agency in the city of Seattle, Washington. She has a large customer base of local department stores, boutiques, discount clothing chain stores, hardware supply catalog stores, and others. A wide variety of models sign yearly contracts with Natasha's agency. In return, she matches models with her customers' needs. Natasha has been very successful, and each year her top-10 models have a chance to audition for a contract with an affiliated (and very prestigious) modeling agency in New York City.

Aunt Natasha knows that you are computer savvy and heard that you are good with database design and Microsoft Access. She employs you to design her database and set up a prototype on Microsoft Access. Later, this prototype will be migrated to a larger database system with Internet access.

Overview of Information Needs

You interview Natasha, and she describes to you the information that the database must contain. First, Natasha needs to keep track of her customers. Just about all her customers are "repeat business," and she wants to make sure that she maintains contact with them. Some customers might hire models infrequently, but when they do, the modeling agency can earn substantial fees. For example, a clothing company that mails out twice-yearly catalogues might employ 20 of Natasha's models for a week-long photo shoot.

As part of a customer's basic information, the name of a contact person should be recorded. In addition to that basic information, a customer's information should note whether the customer will reimburse a model for travel expenses (some do not). This is important because some models are located out of the area and might not accept a job if they must cover their own travel expenses.

Basic contact information for each model must also be recorded. In addition, each model's date of birth must be noted because some models are under 18 years of age and must have their legal guardian sign a release form. An affirmation of that signature must be included in the data. Further, each model's "type" should be noted. All models are categorized by type, for example, *child model*, *general fashion model*, *plus-size model*, etc. Finally, at least one photo of each model should be included.

Each booking must also be recorded. Booking information should show which company is hiring which model, on which date, for how long, and for what fee.

Database Forms

One very important aspect of the database will be the forms, which eventually will be migrated to the Internet. The first form is for model information. People who want to be agency models can use the form to register with the agency. Later, they might sign a contract with the modeling agency. The form should include information such as the model's type, name, address, and phone number. The form should also include a space for the model's photograph. (A photo of the model is crucial!)

The second form will be model-request form for customers. The form should gather basic customer information, the dates and times for which models are needed, and how much the customer is willing to pay. (Natasha's top models are rather expensive, but those who are just entering the business are willing to work for much less because they want the experience.)

Queries

Natasha wants you to create the five queries that follow. In time, she will probably want to add additional queries.

1. When customers call the agency, they always ask for models of a certain type. Thus, when Natasha gets a telephone call or a customer request via the Internet form, she would like a query that prompts for a type of model. Agency personnel should be able to type in the specified model type, such as *child model*, and then receive a list of all the models in that category.
2. Another query needs to show the number of bookings for each model, from top number of bookings to lowest number of bookings. This information will eventually be used to select the top-10 models to audition for work in the affiliated agency in New York City. Natasha gets a very generous bonus for successful referrals.
3. Natasha wants to "grow the business," and she wants to ensure that she recruits enough of the most popular model types to satisfy demand. Thus, she needs to be able to run a query that provides a list of the number of model bookings by model type. This query should list model types from the most popular model type to the least popular model type.
4. Models also ask Natasha for information. Some models live quite a distance from the area, and before accepting a booking, they want to know whether their travel will be paid. Natasha needs to run a query that will quickly answer that question.
5. A query needs to be run that will allow Natasha to update customers' fees. For example, the top model has threatened to quit and move to a different agency (his contract is up soon) if he doesn't get a raise. Natasha has spoken to her customers, and they have agreed to raise his fee by $50 an hour from a certain date onward. Because he has so many bookings, an update query will be an efficient way to increase all those fees.

Report

The agency needs a report that calculates the agency's fees for each model by job. The agency gets 30% of each model's booking fee. The report should show each model's last name, the customer ID number, the customer's name, the booking date, and the agency's fees, subtotaled for the model.

Switchboard

Finally, a switchboard needs to be created to make an easy entry point for the forms and the report.

Assignment 1 Creating the Database Design

In this assignment, you will design your database tables on paper, and then use a word-processing program to document them. Pay close attention to the tables' logic and structure. Do not start your Access code (Assignment 2) before getting feedback from your instructor on Assignment 1. Keep in mind that you will need to look at what is required in Assignment 2 to design your fields and tables properly. It's good programming practice to look at the required outputs before designing your database. When designing the database, observe the following guidelines:

- First, determine the tables you'll need by listing on paper the name of each table and the fields that it should contain. Avoid data redundancy. Do not create a field if it could be created by a "calculated field" in a query.
- Include a logical field that answers the questions, "Does the company reimburse travel expenses?" and "Has the legal guardian of a model under age 18 signed the contract?"
- You'll need at least one transaction table. Avoid duplicating data.
- Document your tables by using the Table facility of your word processor. Your word-processed tables should resemble the format of the table in Figure 4-1.
- You must mark the appropriate key field(s). You can designate a key field by an asterisk (*) next to the field name. Keep in mind that some tables need a compound primary key to uniquely identify a record within a table.
- Print the database design.

TABLE NAME	
Field Name	*Data Type (text, numeric, currency, etc.)*
...	...
...	...

Figure 4-1 Table design

Case 4

Have your design approved before beginning Assignment 2; otherwise, you may need to redo Assignment 2.

ASSIGNMENT 2 CREATING THE DATABASE, FORMS, QUERIES, AND A REPORT

In this assignment, you will first create database tables in Access and populate them with data. Next, you will create two forms, five queries, a report, and a switchboard.

Assignment 2A: Creating Tables in Access

In this part of the assignment, you will create your tables in Access. Use the following guidelines:

- Type records into the tables, using local stores and businesses as the customers. Assume that the higher-priced stores and businesses reimburse travel expenses. Create at least eight customers.
- Invent and then type the names of different models. Gather at least 10 records of data. If you have a digital camera, take photos of friends and use those for data. If you do not, use clip art for photos. (*Hint*: Make sure the field is the correct data type to accept photos.)
- Use the following types of models: *high fashion*, *general fashion*, *petite*, *plus size*, *swimsuit*, *child*, *commercial print*, and *disabled*.
- Type at least 20 bookings. Assume that booking fees have a $500 minimum per day.

- Appropriately limit the size of the text fields; for example, a zip code does not need to be the default setting of 50 characters in length.
- Print all tables.

Assignment 2B: Creating Forms, Queries, a Report, and a Switchboard

There are two forms, five queries, one report, and one switchboard to generate, as outlined in the background of this case.

Form 1: Models

Create a form for prospective and existing models to enter their personal information. Include a space for a photograph. This form should be based on one table only. Call the form Models. Print one record from that form.

Form 2: Request Booking

Create a form and subform for customers to request their bookings. All information about the customer should be included in the main form. The subform should record all the bookings for that customer. View information and bookings for one customer and print that one record. Call the form Request Booking.

Query 1: Models by Type

Create a query called Models by Type. Assume that customers telephone or fax and ask for a list of models by model type. In output, column headers should display the Model Last Name, Model First Name, Address, City, State, Zip, and Phone when prompted for a particular model type. If you input *Child Model* when prompted, your output format should resemble that shown in Figure 4-2, but with different data.

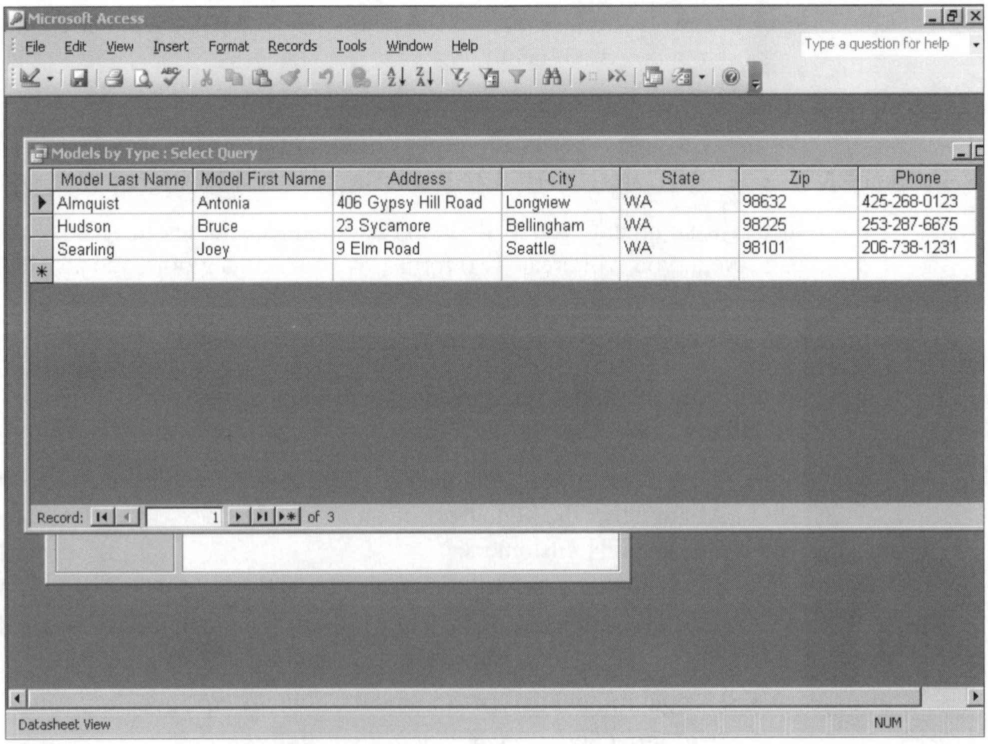

Model Last Name	Model First Name	Address	City	State	Zip	Phone
Almquist	Antonia	406 Gypsy Hill Road	Longview	WA	98632	425-268-0123
Hudson	Bruce	23 Sycamore	Bellingham	WA	98225	253-287-6675
Searling	Joey	9 Elm Road	Seattle	WA	98101	206-738-1231

Figure 4-2 Query 1: Models by Type

Query 2: Most Requested Models

Create a query called Most Requested Models. Your output column headers should be Model ID, Model Last Name, Model First Name, and Number of Bookings (that the model has had so far this year). To simplify things, assume that your data is only from this year. List the most-booked model to the least-booked model. Make sure all column headings are visible and accurate. Your data will differ, but your output format should look like that shown in Figure 4-3.

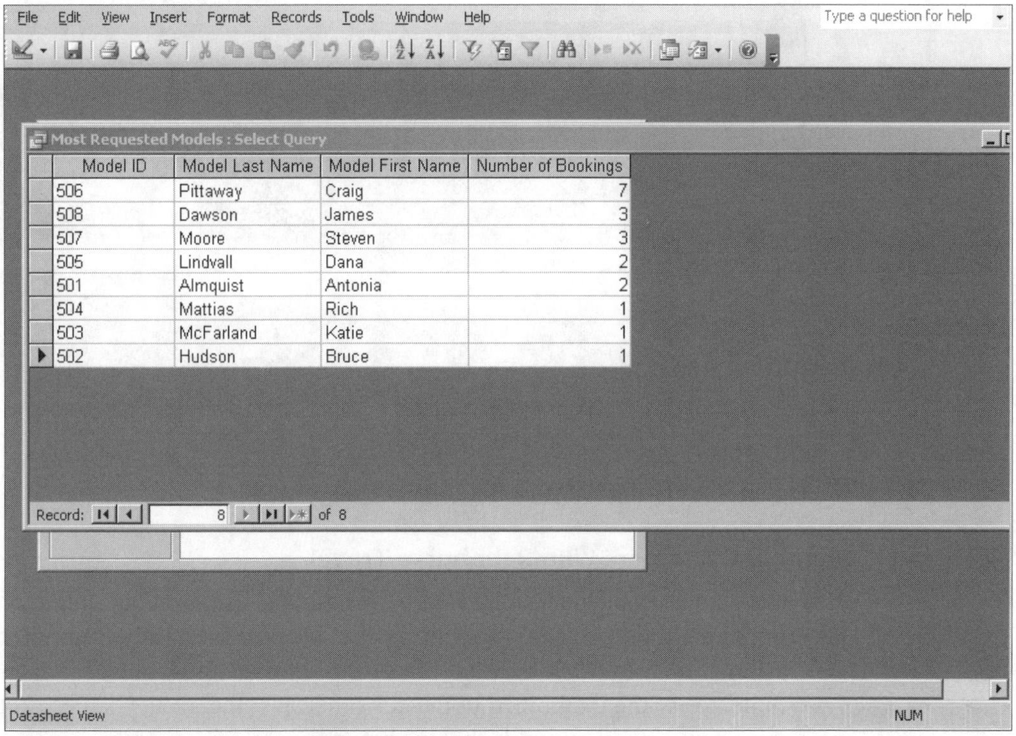

Model ID	Model Last Name	Model First Name	Number of Bookings
506	Pittaway	Craig	7
508	Dawson	James	3
507	Moore	Steven	3
505	Lindvall	Dana	2
501	Almquist	Antonia	2
504	Mattias	Rich	1
503	McFarland	Katie	1
502	Hudson	Bruce	1

Figure 4-3 Query 2: Most Requested Models

Query 3: Most Requested Type of Model

Create a query called Most Requested Type of Model. This is very similar to Query 2. Your output column headers should be Model Type, Model Description, and Number of Bookings (that this model type has had so far this year). As before, assume that your data is only from this year. List the most-booked model type to the least-booked model type. Make sure all column headings are visible and accurate. Note that the report will not show the total billed to the customer, just the agency's fees. Your data will differ, but your output format should look like that shown in Figure 4-4.

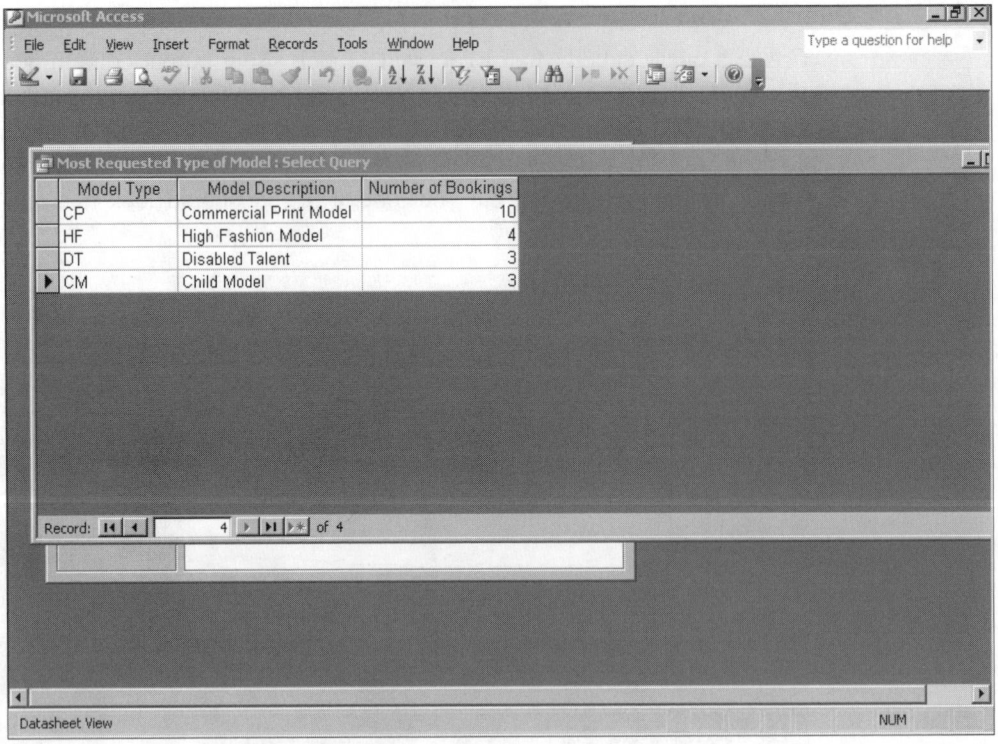

Model Type	Model Description	Number of Bookings
CP	Commercial Print Model	10
HF	High Fashion Model	4
DT	Disabled Talent	3
CM	Child Model	3

Figure 4-4 Query 3: Most Requested Type of Model

Query 4: Customers Who Reimburse Travel

Create a query called Customers Who Reimburse Travel. Your output should show only those companies that will pay models for travel expenses. Your output headers should show Customer Name, Address, City, State, Zip, Phone, and Contact Person. Note that the report will not show the total billed to the customer, just the agency's fees. Your data will differ, but your output should resemble that in shown Figure 4-5.

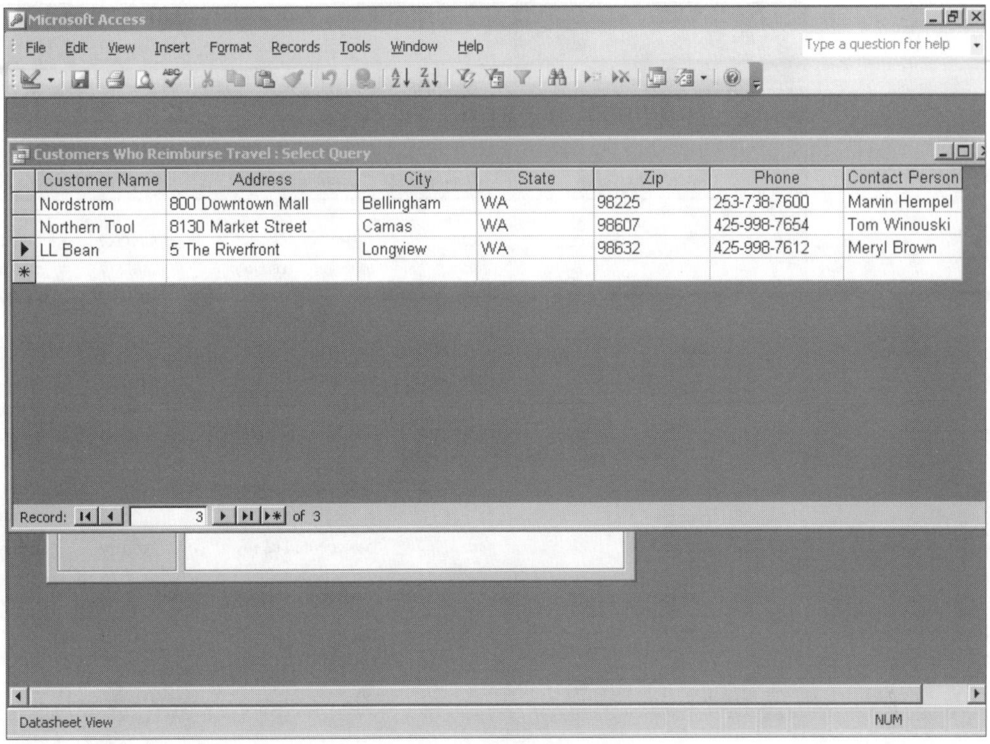
Microsoft Access
File Edit View Insert Format Records Tools Window Help
Type a question for help
Customers Who Reimburse Travel : Select Query
Customer Name | Address | City | State | Zip | Phone | Contact Person
Nordstrom | 800 Downtown Mall | Bellingham | WA | 98225 | 253-738-7600 | Marvin Hempel
Northern Tool | 8130 Market Street | Camas | WA | 98607 | 425-998-7654 | Tom Winouski
LL Bean | 5 The Riverfront | Longview | WA | 98632 | 425-998-7612 | Meryl Brown
Record: 3 of 3
Datasheet View
NUM

Figure 4-5 Query 4: Customers Who Reimburse Travel

Query 5: Increased Fees

Create a query called Increased Fees. This query increases the fee for the top model by $50 for each booking after a certain date, such as 5/1/07. Check with the Most Requested Models query output to see who is the most requested model. Run the query and save it.

Report: Summary of Agency's Fees

Create a report titled Summary of Agency's Fees. Your report's output should show headers for Model Last Name, Customer ID, Customer Name, Date, and Agency's Fee.

First, create a query for input to that report, including calculating the 30% agency's fee. Bring that query into the Report Wizard. Group the report on Model Last Name and subtotal each model's Agency's Fees. Note that the report will not show the total billed to the customer, just the agency's fees. Your data will differ, but your output should resemble that shown in Figure 4-6.

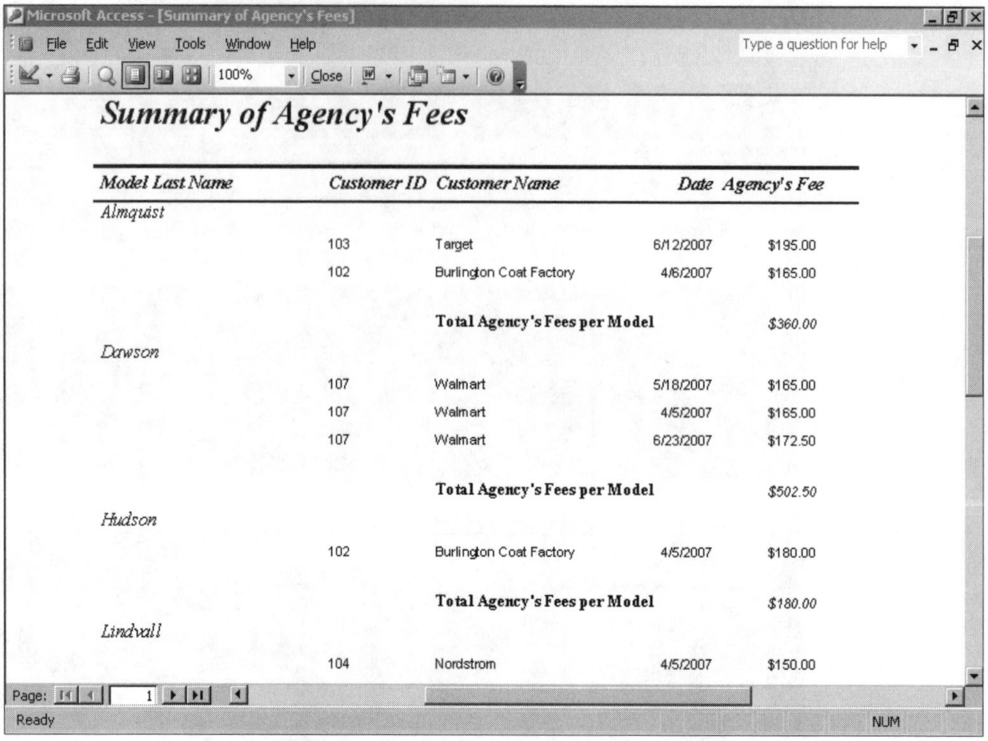

Summary of Agency's Fees

Model Last Name	Customer ID	Customer Name	Date	Agency's Fee
Almquist				
	103	Target	6/12/2007	$195.00
	102	Burlington Coat Factory	4/6/2007	$165.00
		Total Agency's Fees per Model		*$360.00*
Dawson				
	107	Walmart	5/18/2007	$165.00
	107	Walmart	4/5/2007	$165.00
	107	Walmart	6/23/2007	$172.50
		Total Agency's Fees per Model		*$502.50*
Hudson				
	102	Burlington Coat Factory	4/5/2007	$180.00
		Total Agency's Fees per Model		*$180.00*
Lindvall				
	104	Nordstrom	4/5/2007	$150.00

Figure 4-6 Report: Summary of Agency's Fees

Switchboard: The Modeling Agency

Create a switchboard to access the forms and report easily. Call your switchboard The Modeling Agency. Your switchboard should look like that shown in Figure 4-7.

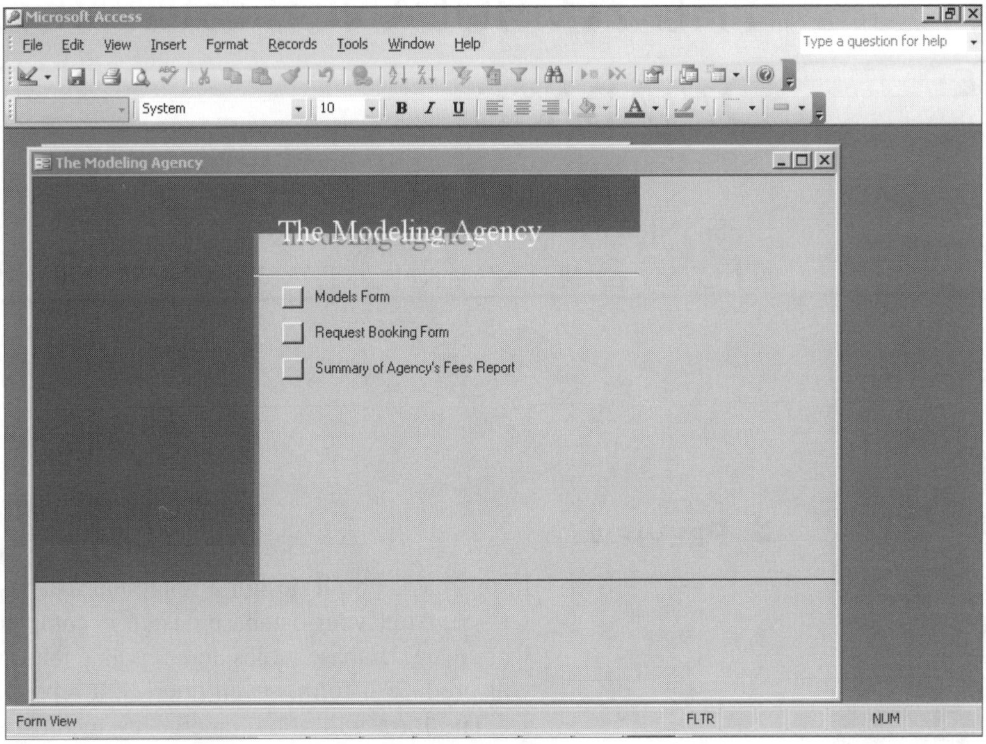

Figure 4-7 Switchboard: The Modeling Agency

ASSIGNMENT 3 MAKING A PRESENTATION

Create a presentation for the modeling agency. Describe your database design and show the design of your tables using PowerPoint. Then demonstrate the use of the forms, queries, report, and switchboard. Discuss future improvements and the migration to the Internet. Your presentation should take fewer than 15 minutes, including a brief question-and-answer period.

DELIVERABLES

1. Word-processed design of tables
2. Tables created in Access
3. Form 1: Models
4. Form 2: Request Booking
5. Query 1: Models by Type
6. Query 2: Customers Who Reimburse Travel
7. Query 4: Most Requested Models
8. Query 5: Most Requested Type of Model
9. Report: Summary of Agency's Fees
10. Presentation materials
11. Any other required tutorial printouts or tutorial CD

Staple all pages together. Put your name and class number at the top of the page. Make sure your disk or CD is labeled, if required.

5 CASE

The Ambulance Service Database

Designing a Relational Database to Create Tables, a Form, Queries, a Report, and a Switchboard

Preview

In this case, you'll design a relational database for an ambulance service. After your database design is completed and correct, you will create database tables and populate them with data. Then you will produce a form, seven queries, a report, and a switchboard.

The form will allow emergency medical technicians (EMTs) to clock in and out of their shifts. The queries will calculate the frequency of calls to each zip code, for each type of emergency, and by each base station. Other queries will calculate the time it takes to respond to heart-attack calls, the qualifications of the EMTs compared to the types of emergencies to which they respond, and which EMTs are on a particular call. An Update query will be created to change the qualification status of some EMTs. The report will summarize the ambulance calls by zip code. The switchboard will allow ambulance service employees to record their hours and view the report.

Preparation

- Before attempting this case, you should have some experience in database design and in using Microsoft Access.
- Complete any part of Database Design Tutorial A that your instructor assigns.
- Complete any part of Access Tutorial B that your instructor assigns, or refer to the tutorial as necessary.
- Refer to Tutorial E as necessary.

Background

Three unprofitable ambulance companies in a small township have merged in order to consolidate operations and lower costs. The three companies are located in different areas of the township, and historically they have had some overlap of coverage. However, the new ambulance service has received numerous complaints from people in the township. These complaints range from slow service to poorly trained EMTs. In addition, it seems as though the three ambulance base stations have differing call volumes and types of calls. You are hired as the Operations Coordinator, experienced in database design and Microsoft Access, to straighten out the problems of this newly merged service.

Understanding the Problems

There are a number of parameters to keep in mind when designing this database. First, the emergency medical technicians' training varies. There are four levels of training, and each EMT is at one of those levels. An EMT's training level predicates what he or she can do for a patient. This information is important because there have been complaints that some EMTs are not adequately trained to care for patients en route to a hospital. You've noted that one base station seems to cover an urban area that has a high level of violent crime—highly trained EMTs are needed when responding to these calls. By contrast, another area of the township has mostly retired people, and the base station responding to these calls must be prepared to stabilize heart-attack patients.

The database also has to keep track of the different ambulances the company owns and to which base station site a particular ambulance is assigned. Each of the three original companies had its own site and vehicles, and the merged company is still using those same three sites and vehicles. You've observed that some vehicles are better equipped than others. To respond to calls promptly, the ambulance service needs to know which vehicles are available and, in some cases, which might best meet the needs of a particular call.

When the ambulance service receives a call, usually via 911, certain information is recorded: the call time, the time the ambulance arrived on the scene, the type of emergency, and the address with the zip code. The zip code is important for tracking purposes: Knowing which zip codes receive the highest volume of calls will allow you to make sure that base stations are adequately staffed and have sufficient vehicles to meet the needs of that area. In the future, you plan to use geographical information systems (GIS) mapping to overlay the call data and zip code data. With such a profile in hand, you might recommend reassigning personnel and vehicles—or establishing a new base station if there are gaps in coverage.

For payroll purposes, the database needs to include the hours worked by the EMTs. For legal reasons, it's also important to be able to trace which EMT worked on which shift and responded to which call. You'll also be able to use this information to assess personnel needs at each base station.

Some EMTs have complained that they are not being paid for all the hours that they work. EMTs need to clock in and out of their shifts in the ambulance service easily. Creating a form of the EMTs' information, with a subform of their hours worked, will be a convenient way for them to keep track of their hours.

Planning to Use the Database

As operations coordinator, you will need to query the database to understand the situation at the newly merged ambulance service so you can make any needed changes in staffing or vehicle garaging. You'll begin by setting up a series of queries:

1. First, you'll need a record of the frequency of calls by types of emergency: heart attack, vehicle accident, gunshot wounds, etc. This will give you an idea of the different types of emergencies the service encounters at its various locations.
2. Monitoring the frequency of calls for each base station is vital to managing the operation. As mentioned, there are three sites, and some have more calls than the others. You can help to eliminate the complaint of slow service by having each base station adequately staffed. Running a daily query of call volumes will provide a useful profile.
3. In addition, every day you will need to list the frequency of calls in all the zip codes the three base stations service. Each base station services multiple zip codes. The original three sites of the ambulances might need to be moved—or a new base station added—to accommodate the volume of calls in certain zip codes.
4. You are also interested in monitoring the time that it takes to respond to heart-attack calls. A slow response time can be the difference between life and death. Your query will figure the elapsed time it takes to get to the emergency when it's a heart attack.
5. You also want to gauge whether EMTs are adequately trained for the type of emergency calls to which they respond. You want to query the database to compare the kind of emergency tracked against the EMTs' qualification and training. This information will be for staffing needs and re-qualification proposals. Thus, by comparing the number of calls for a particular emergency against the number of EMTs in each of the four basic training categories (First Responder EMTs, Basic EMTs, Intermediate EMTs, and Paramedics), you can decide whether more highly qualified EMTs are needed. This will be done for each base station , so the query needs to prompt for the name of a base. Most vehicles contain a number of EMTs, and the training of those EMTs will vary in each vehicle. This query will give management a rough idea of whether that spread of differing qualified EMTs is balanced in each base station.
6. For several reasons, you also need to identify which EMTs responded to each call. Sometimes, emergency victims telephone the ambulance service after they have recovered and want to thank the EMTs that might have saved their lives. Another reason is that you might want to know the training qualifications of the EMTs who have gone out on a particular call, especially if there is a customer complaint or a liability situation.
7. After sorting through all your information, you offer to pay for training to take all the EMTs at Level 2 up to Level 3. You need to create an update query that will change their qualification status after the training is complete.

A report summarizing all the calls to the ambulance service by zip code in a given period of time is also necessary for the management of the company. By looking at the zip codes, dates, call times, and types of emergency, you can plan for changes in staffing, equipping your vehicles, and location of vehicles at peak times.

The switchboard will be a convenient way for those EMTs not familiar with Microsoft Access to record their hours worked and the vehicles in which they rode.

Assignment 1 Creating the Database Design

In this assignment, you will design your database tables on paper, using a word-processing program. Pay close attention to the tables' logic and structure. Do not start your Access code (Assignment 2) before getting feedback from your instructor on Assignment 1. Keep in mind that you will need to look at what is required in Assignment 2 to design your fields and tables properly. It's good programming practice to look at the required outputs before designing your database. When designing the database, observe the following guidelines:

- First, determine the tables you'll need by listing on paper the name of each table and the fields that it should contain. Avoid data redundancy. Do not create a field if it could be created by a "calculated field" in a query.
- You'll need at least one transaction table. Avoid duplicating data.
- Document your tables by using the Table facility of your word processor. Your word-processed tables should resemble the format of the table in Figure 5-1.
- You must mark the appropriate key field(s). You can designate a key field by an asterisk (*) next to the field name. Keep in mind that some tables might need a compound primary key to uniquely identify a record within a table.
- Print out the database design.

TABLE NAME	
Field Name	*Data Type (text, numeric, currency, etc.)*
...	...
...	...

Figure 5-1 Table design

Have your design approved before beginning Assignment 2; otherwise, you may need to redo Assignment 2.

Assignment 2 Creating the Database and a Form, Queries, a Report, and a Switchboard

In this assignment, you will first create database tables in Access and populate them with data. Next, you will create a form, seven queries, a report, and a switchboard.

Assignment 2A: Creating Tables in Access

In this part of the assignment, you will create your tables in Access. Use the following guidelines:

- Type records into the tables, using your classmates' names as those of the EMTs. Create at least 10 records, and have each EMT be rated at Level 1, 2, 3, or 4. Invent address and telephone data.
- Assume that Level 1 (least training) is called a First Responder, Level 2 is Basic, Level 3 is Intermediate, and Level 4 is Paramedic.

- Create at least five vehicles, each one being garaged at one of three base stations. Assign each vehicle a unique ID and a town location. To minimize typing, we will assume each vehicle is a standard ambulance vehicle.
- Make up three base station names.
- Create at least nine calls on one particular date. Make each "type of emergency" one of the following: *childbirth*, *auto accident*, *drowning*, *heart attack*, or *gunshot wound*. Assign each emergency a unique ID, and have the location of the emergency a street address and zip code only. Have the emergencies in an area containing four zip codes.
- Record the EMTs' names who are working during each of the calls you created.
- Appropriately limit the size of the text fields; for example, a zip code does not need to be the default setting of 50 characters in length.
- Print all tables.

Assignment 2B: Creating a Form, Queries, a Report, and a Switchboard

There is one form, seven queries, one report, and one switchboard to create, as outlined in the background of this case.

Form: EMT Work Hours

Create a form called EMT Work Hours. Include all the information on the EMTs. Using a subform, include all the information on the work hours of the EMTs, such as the Date, Vehicle, Starting Hour, and Ending Hour.

Query 1: Frequency of Calls by Type of Emergency

Create a query called Frequency of Calls by Type of Emergency. The column headers should be for Type of Emergency, the Number of Calls, and the Base station name. Adjust the column headings as needed. Sort the data so that the most frequent emergency is at the top of the list. Your output format should resemble that shown in Figure 5-2, but with different data.

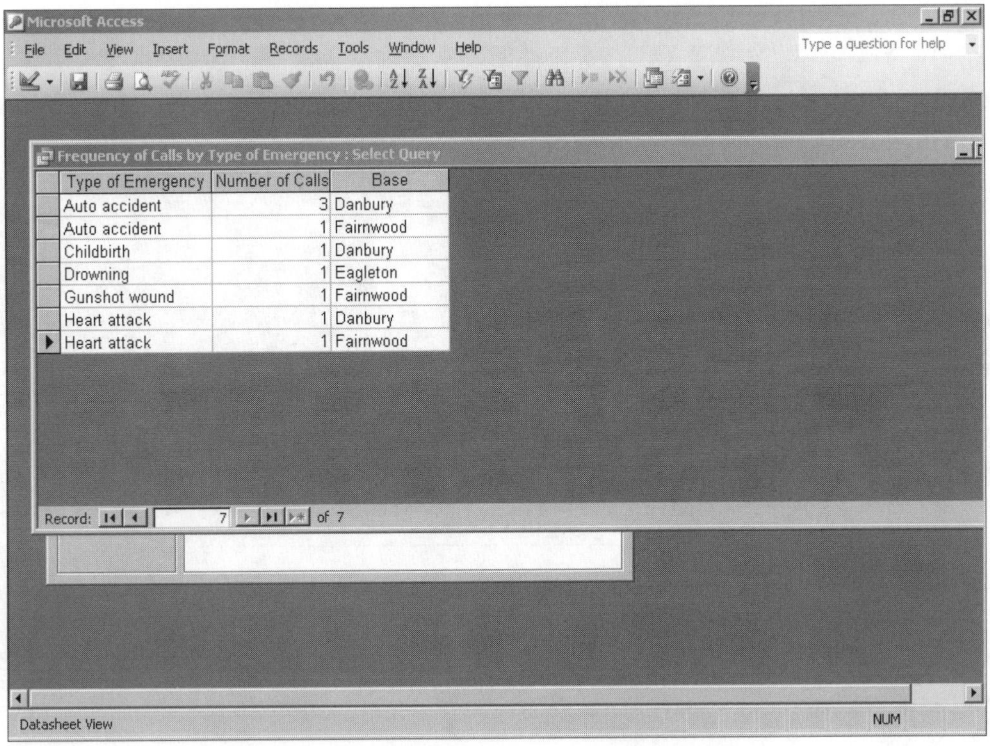
Microsoft Access
File Edit View Insert Format Records Tools Window Help
Type a question for help
Frequency of Calls by Type of Emergency : Select Query
Type of Emergency | Number of Calls | Base
Auto accident | 3 | Danbury
Auto accident | 1 | Fairnwood
Childbirth | 1 | Danbury
Drowning | 1 | Eagleton
Gunshot wound | 1 | Fairnwood
Heart attack | 1 | Danbury
Heart attack | 1 | Fairnwood
Record: 7 of 7
Datasheet View
NUM

Figure 5-2 Query 1: Frequency of Calls by Type of Emergency

Query 2: Frequency of Calls by Base

Create a query called Frequency of Calls by Base. This query is identical to Query 1 except that you will display the number of calls and the base station, listing the most frequently called base station first. Your data will differ, but your output format should resemble that shown in Figure 5-3.

Number of Calls	Base
5	Danbury
3	Fairnwood
1	Eagleton

Figure 5-3 Query 2: Frequency of Calls by Base

Query 3: Frequency of Calls by Zip

Create a query called Frequency of Calls by Zip. Again, the directions are similar to the last two queries except for displaying the zip code and the number of calls. Sorting should show the most frequently called zip code at the top of the list. Your data will differ, but the output format should resemble that shown in Figure 5-4.

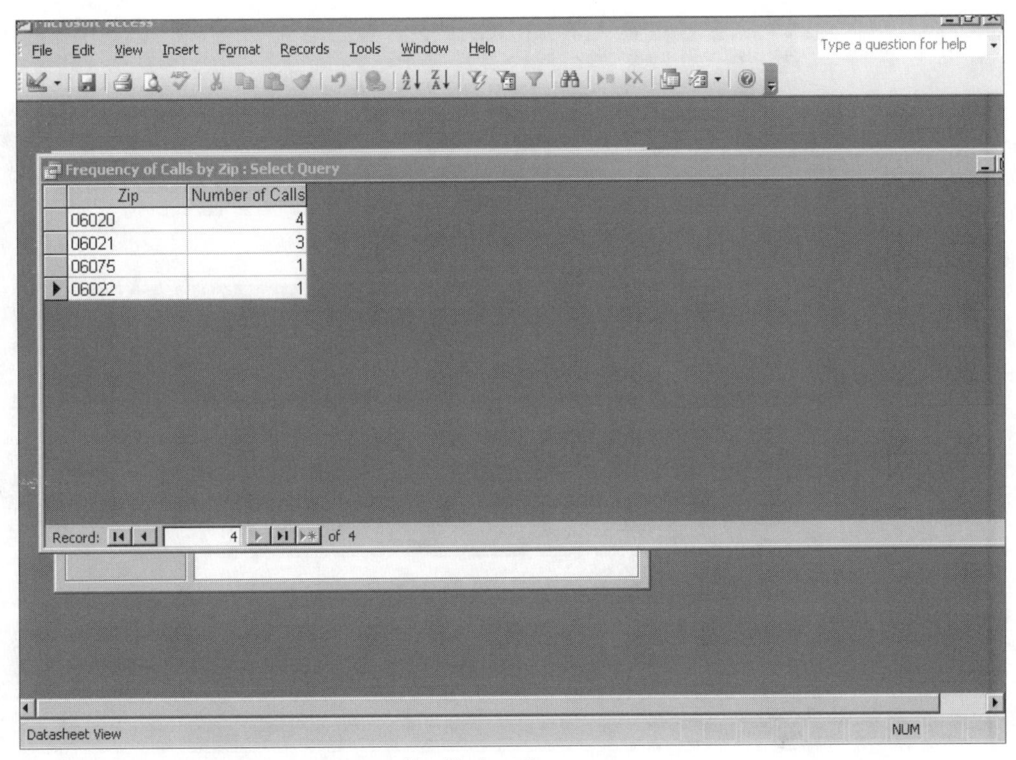

Figure 5-4 Query 3: Frequency of Calls by Zip

Case 5

Query 4: Time to Respond to Heart Attacks

Create a query called Time to Respond to Heart Attacks. Your output should show the column headers: Call #, Date, Vehicle ID, Type of Emergency, and Elapsed Time (the time it takes to get to the emergency). Your data will differ, but your output format should look like that shown in Figure 5-5.

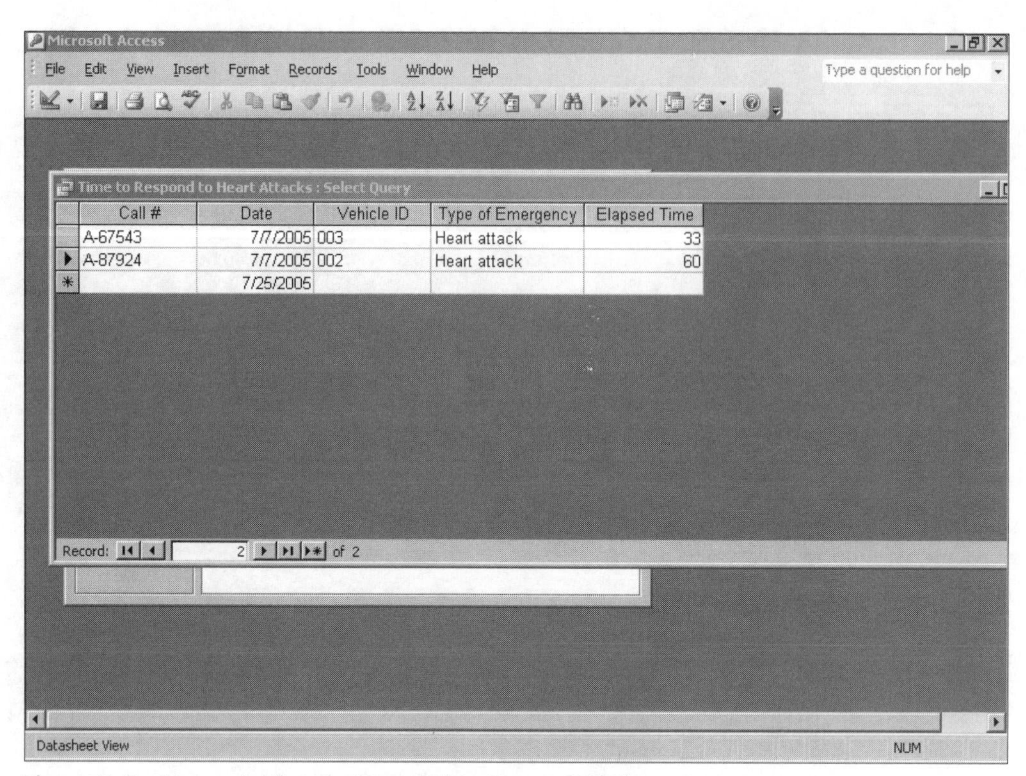

Figure 5-5 Query 4: Time to Respond to Heart Attacks

Query 5: Types of Emergencies and Qualifications

Create a query called Types of Emergencies and Qualifications. This query should prompt for a particular base. Your output should have these headers: Type of Emergency, Qualification and Training, the Number of Calls, and Base. Sort on the Type of Emergency. Your data will differ, but your output format should look like that shown in Figure 5-6, if you were to input the base station name Danbury.

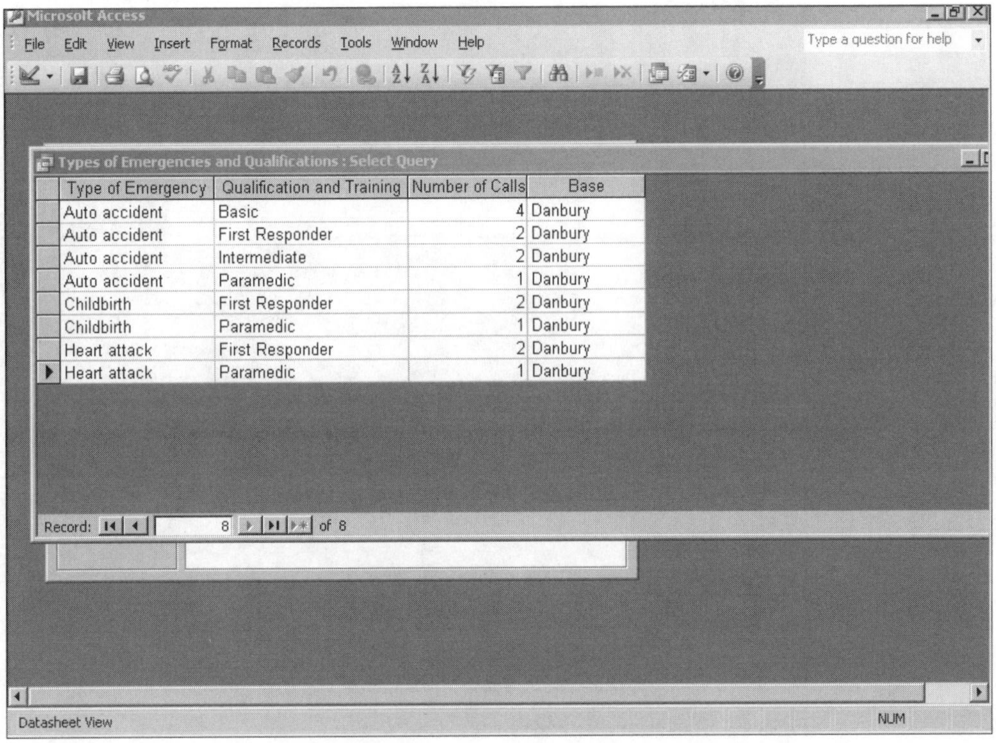

Type of Emergency	Qualification and Training	Number of Calls	Base
Auto accident	Basic	4	Danbury
Auto accident	First Responder	2	Danbury
Auto accident	Intermediate	2	Danbury
Auto accident	Paramedic	1	Danbury
Childbirth	First Responder	2	Danbury
Childbirth	Paramedic	1	Danbury
Heart attack	First Responder	2	Danbury
Heart attack	Paramedic	1	Danbury

Figure 5-6 Query 5: Types of Emergencies and Qualifications

Query 6: Re-Qualification of EMTs

Create an Update query called Re-Qualification of EMTs. The query should change all the qualification levels of EMTs whose current qualification is Level 2 (Basic EMT) to Level 3 (Intermediate EMT). Save the query and run it.

Query 7: EMTs on a Call

Create a query called EMTs on a Call #. This query should prompt for a particular call number and then display the Type of Emergency, and EMT information: employee's Last Name, First Name, and Qualification and Training. Your data will differ, but when you input a call number, a portion of your output should resemble that shown in Figure 5-7.

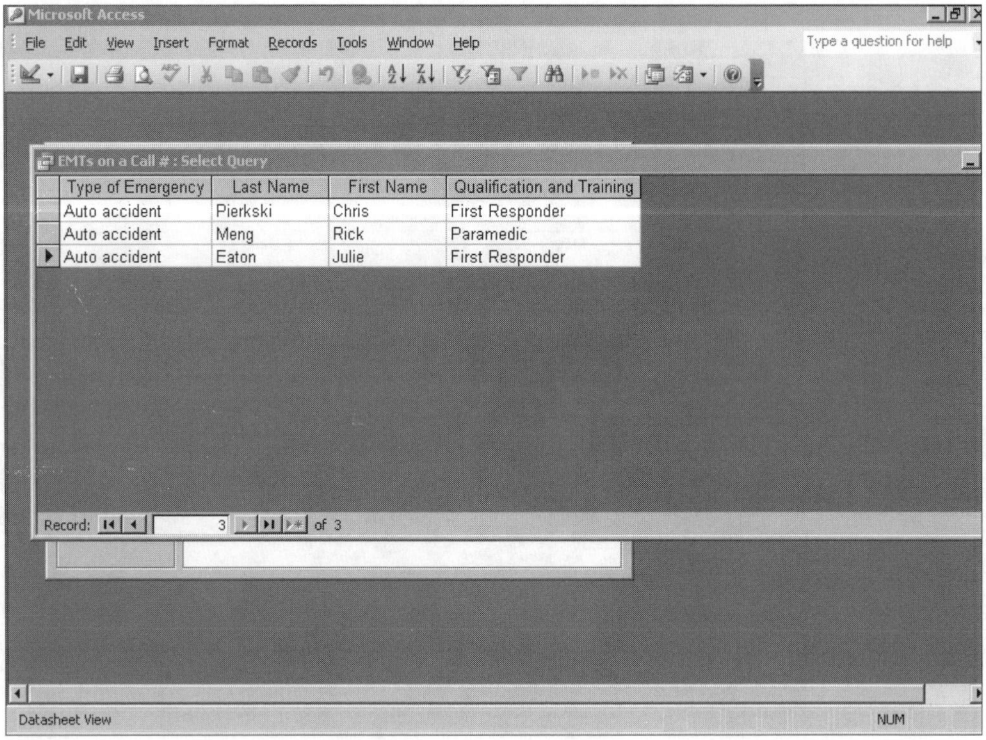

Type of Emergency	Last Name	First Name	Qualification and Training
Auto accident	Pierkski	Chris	First Responder
Auto accident	Meng	Rick	Paramedic
Auto accident	Eaton	Julie	First Responder

Figure 5-7 Query 7: EMTs on a Call #

Report: Emergency Call Summary

Create a report titled Emergency Call Summary. Your report's output should show headers for Zip, Date, Call Time, and Type of Emergency. Group the report on the zip code. Your data will differ, but your output format should resemble that shown in Figure 5-8.

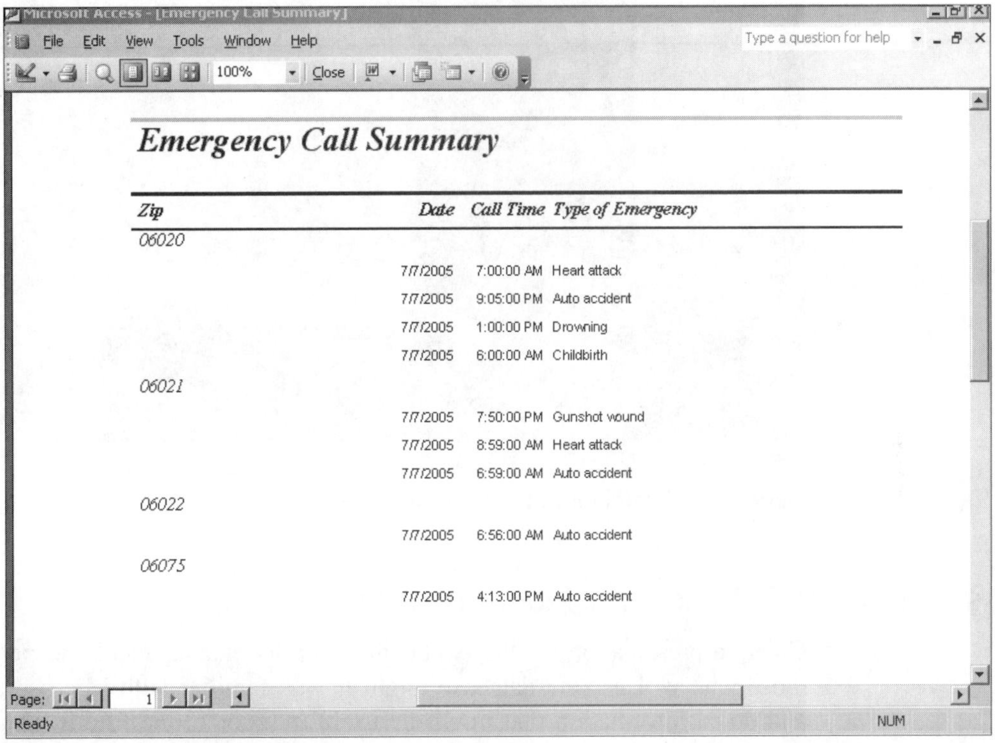

Microsoft Access - [Emergency Call Summary]

File Edit View Tools Window Help   Type a question for help

100% Close

Emergency Call Summary

Zip	*Date*	*Call Time*	*Type of Emergency*
06020			
	7/7/2005	7:00:00 AM	Heart attack
	7/7/2005	9:05:00 PM	Auto accident
	7/7/2005	1:00:00 PM	Drowning
	7/7/2005	6:00:00 AM	Childbirth
06021			
	7/7/2005	7:50:00 PM	Gunshot wound
	7/7/2005	8:59:00 AM	Heart attack
	7/7/2005	6:59:00 AM	Auto accident
06022			
	7/7/2005	6:56:00 AM	Auto accident
06075			
	7/7/2005	4:13:00 PM	Auto accident

Page: 1

Ready   NUM

Figure 5-8 Report: Emergency Call Summary

Switchboard: The Ambulance Service

Create a switchboard to access the form and report easily. Call your switchboard The Ambulance Service. Your switchboard should look like that shown in Figure 5-9.

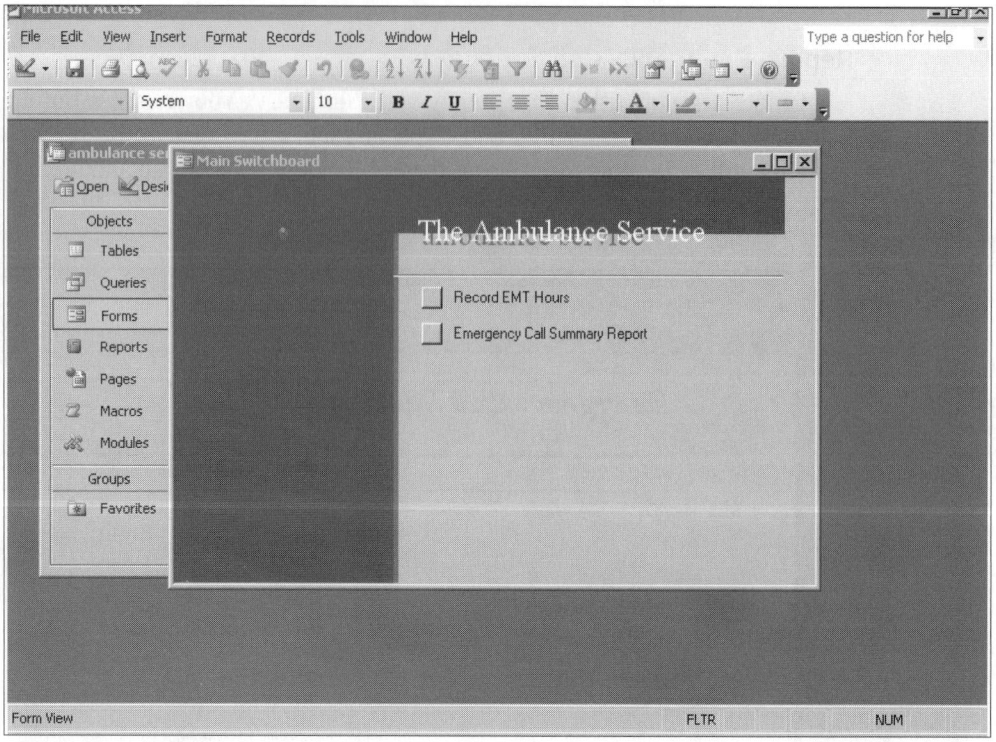

Figure 5-9 Switchboard: The Ambulance Service

ASSIGNMENT 3 MAKING A PRESENTATION

Create a presentation for the Ambulance Service management and personnel. Consider that those who will use the database might not be familiar with Microsoft Access. Present any additional information that might be useful now or in the future. Your presentation should take fewer than 15 minutes, including a brief question-and-answer period.

DELIVERABLES

1. Word-processed design of tables
2. Tables created in Access
3. Form: EMT Work Hours
4. Query 1: Frequency of Calls by Types of Emergencies
5. Query 2: Frequency of Calls by Base
6. Query 3: Frequency of Calls by Zip
7. Query 4: Time to Respond to Heart Attacks
8. Query 5: Types of Emergencies and Qualifications
9. Query 7: EMTs on a Call #
10. Report: Emergency Call Summary
11. Presentation materials
12. Any other required tutorial printouts or tutorial CD

Staple all pages together. Put your name and class number at the top of the page. Make sure your disk or CD is labeled, if required.

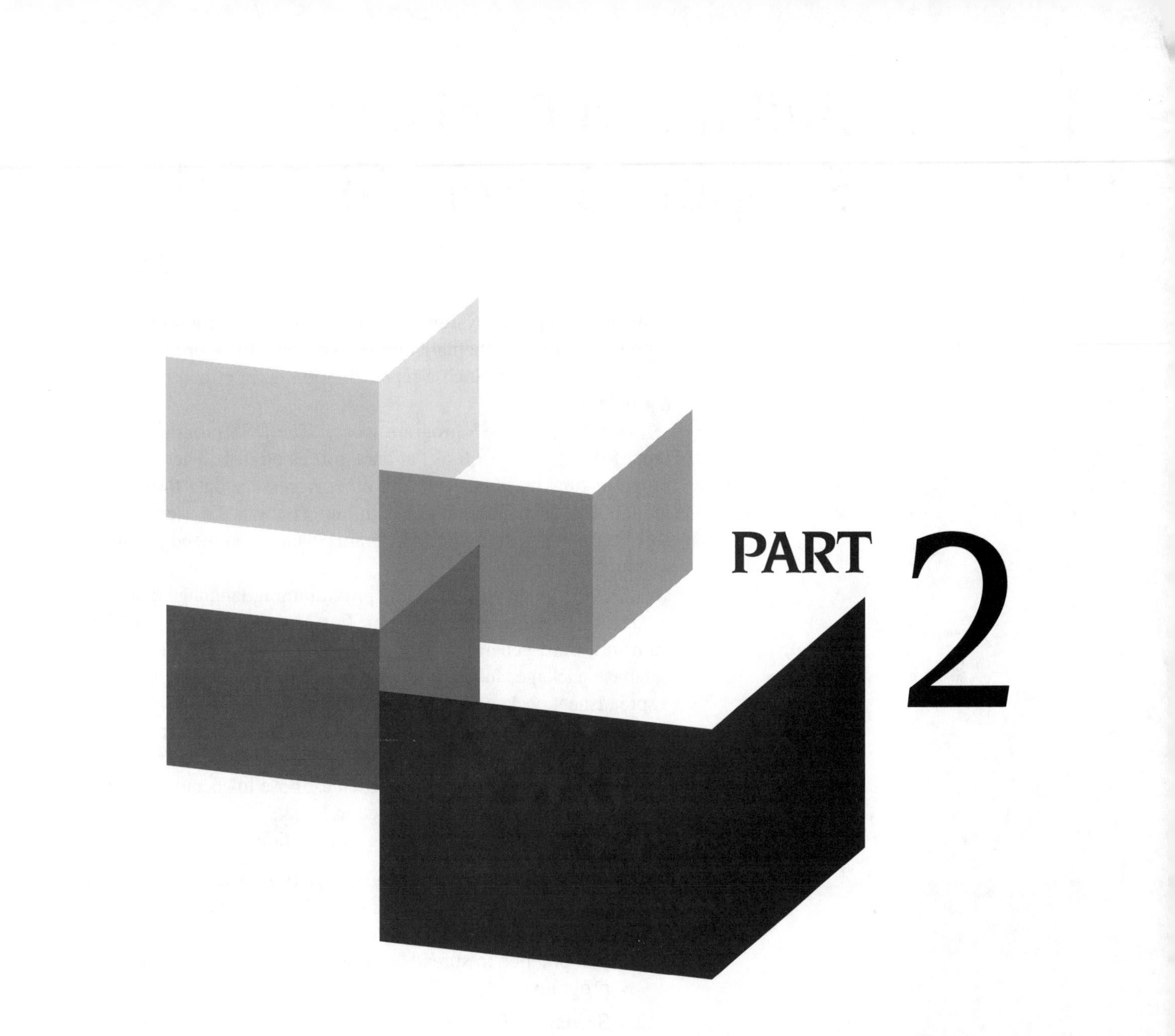

Decision Support Cases Using Excel Scenario Manager

C
TUTORIAL

Building a Decision Support System in Excel

A **decision support system (DSS)** is a computer program that can represent, either mathematically or symbolically, a problem that a user needs to solve. Such a representation is, in effect, a model of a problem.

Here's how a DSS program works: The DSS program accepts input from the user or looks at data in files on disk. Then, the DSS program runs the input and any other necessary data through the model. The program's output is the information the user needs to solve a problem. Some DSS programs even recommend a solution to a problem.

A DSS can be written in any programming language that lets a programmer represent a problem. For example, a DSS could be built in a third-generation language, such as Visual Basic, or in a database package, such as Access. A DSS could also be written in a spreadsheet package, such as Excel.

The Excel spreadsheet package has standard built-in arithmetic functions as well as many statistical and financial functions. Thus, many kinds of problems—such as those in accounting, operations, or finance—can be modeled in Excel.

This tutorial has the following four sections:

1. **Spreadsheet and DSS Basics:** In this section, you'll learn how to create a DSS program in Excel. Your program will be in the form of a cash flow model. This section will give you practice in spreadsheet design and in building a DSS program.
2. **Scenario Manager:** In this section, you'll learn how to use an Excel tool called the Scenario Manager. With any DSS package, one problem with playing "what if" is this: Where do you physically record the results from running each set of data? Typically, a user just writes the inputs and related results on a piece of paper. Then—ridiculously enough—the user might have to input the data *back* into a spreadsheet for further analysis! The Scenario Manager solves that problem. It can be set up to capture inputs and results as "scenarios," which are then nicely summarized on a separate sheet in the Excel workbook.
3. **Practice Using Scenario Manager:** You will work on a new problem, a case using the Scenario Manager.
4. **Review of Excel Basics:** This brief section reviews the information you'll need to do the spreadsheet cases that follow this tutorial.

SPREADSHEET AND DSS BASICS

Assume it is late in 2006, and that you are trying to build a model of what a company's net income (profit) and cash flow will be in the next two years (2007 and 2008). This is the problem: to forecast net income and cash flow in those years. Assume that knowing these forecasts would help answer some question or make some decision. After researching the problem, you decide that the estimates should be based on three things: (1) 2006 results, (2) estimates of the underlying economy, and (3) the cost of products the company sells.

The model will use an income statement and cash flow framework. The user can input values for two possible states of the economy in years 2007–2008: an "O" for an Optimistic outlook or a "P" for a Pessimistic outlook. The state of the economy is expected to affect the number of units the company can sell as well as the unit selling price: In a good "O" economy, more units can be sold at a higher price. The user can also input values for two possible cost-of-goods-sold price directions: a "U" for Up or a "D" for Down. A "U" means that the cost of an item sold will be higher than it was in 2006; a "D" means that it will be less.

Presumably, the company will do better in a good economy and with lower input costs—but how much better? The relationships are too complex to assess in one's head, but the software model can easily assess the relationships. Thus, the user can play "what if" with the input variables and note the effect on net income and year-end cash levels. For example, a user can ask, "What if the economy is good and costs go up? What will net income and cash flow be in that case? What would happen if the economy is down and costs go down? What would be the company's net income and cash flow in that case?" With an Excel software model available, the answers are easily quantified.

Organization of the DSS Model

Your spreadsheets should have the following sections, which will be noted in boldface type throughout this tutorial and in the Excel cases that follow it:

- **CONSTANTS**
- **INPUTS**
- **SUMMARY OF KEY RESULTS**
- **CALCULATIONS** (of values that will be used in the INCOME STATEMENT AND CASH FLOW STATEMENT)
- **INCOME STATEMENT AND CASH FLOW STATEMENT**

Here, as an extended illustration, a DSS model is built for the forecasting problem previously described. Let's look at each spreadsheet section. Figures C-1 and C-2 show how to set up the spreadsheet.

	A	B	C	D
1	TUTORIAL EXERCISE			
2				
3	CONSTANTS	2006	2007	2008
4	TAX RATE	NA	0.33	0.35
5	NUMBER OF BUSINESS DAYS	NA	300	300
6				
7	INPUTS	2006	2007	2008
8	ECONOMIC OUTLOOK (O = OPTIMISTIC; P = PESSIMISTIC)	NA		NA
9	PURCHASE-PRICE OUTLOOK (U = UP; D = DOWN)	NA		NA
10				
11	SUMMARY OF KEY RESULTS	2006	2007	2008
12	NET INCOME AFTER TAXES	NA		
13	END-OF-THE-YEAR CASH ON HAND	NA		
14				
15	CALCULATIONS	2006	2007	2008
16	NUMBER OF UNITS SOLD IN A DAY	1000		
17	SELLING PRICE PER UNIT	7.00		
18	COST OF GOODS SOLD PER UNIT	3.00		
19	NUMBER OF UNITS SOLD IN A YEAR	NA		

Figure C-1 Tutorial skeleton 1

	A	B	C	D
21	INCOME STATEMENT AND CASH FLOW STATEMENT	2006	2007	2008
22	BEGINNING-OF-THE-YEAR CASH ON HAND	NA		
23				
24	SALES (REVENUE)	NA		
25	COST OF GOODS SOLD	NA		
26	INCOME BEFORE TAXES	NA		
27	INCOME TAX EXPENSE	NA		
28	NET INCOME AFTER TAXES	NA		
29				
30	END-OF-THE-YEAR CASH ON HAND (BEGINNING-OF-THE-YEAR CASH, PLUS NET INCOME AFTER TAXES)	10000		

Figure C-2 Tutorial skeleton 2

Each spreadsheet section is discussed next.

The CONSTANTS Section

This section records values that are used in spreadsheet calculations. In a sense, the constants are inputs, except that they do not change. In this tutorial, constants are TAX RATE and the NUMBER OF BUSINESS DAYS.

The INPUTS Section

The inputs are for the ECONOMIC OUTLOOK and PURCHASE-PRICE OUTLOOK (manufacturing input costs). Inputs could conceivably be entered for *each year* covered by the model (here, 2007 and 2008). This would let you enter an "O" for 2007's economy in one cell and a "P" for 2008's economy in another cell. Alternatively, one input for the two-year period could be entered in one cell. For simplicity, this tutorial uses the *latter* approach.

The SUMMARY OF KEY RESULTS Section

This section will capture 2007 and 2008 NET INCOME AFTER TAXES (profit) for the year and END-OF-THE-YEAR CASH ON HAND, which are (assume) the two relevant outputs of this model. The summary merely repeats, in one easy-to-see place, results that are shown in otherwise widely spaced places in the spreadsheet. This just makes the answers easier to see all at once. (It also makes it easier to graph results later.)

The CALCULATIONS Section

This area is used to compute the following data:

1. The NUMBER OF UNITS SOLD IN A DAY (a function of the 2006 value and of the economic outlook input)
2. The SELLING PRICE PER UNIT (similarly derived)
3. The COST OF GOODS SOLD PER UNIT (a function of the 2006 value and of the purchase-price outlook)
4. The NUMBER OF UNITS SOLD IN A YEAR (equals the number of units sold in a day times the number of business days)

These formulas could be embedded in the **INCOME STATEMENT AND CASH FLOW STATEMENT** section. Doing that would, however, cause the expressions there to be complex and difficult to understand. Putting the intermediate calculations into a separate **CALCULATIONS** section breaks up the work into modules. This is good form because it simplifies your programming.

The INCOME STATEMENT AND CASH FLOW STATEMENT Section

This is the "body" of the spreadsheet. It shows the following:

1. BEGINNING-OF-THE-YEAR CASH ON HAND, which equals cash at the end of the *prior* year.
2. SALES (REVENUE), which equals the units sold in the year times the unit selling price.
3. COST OF GOODS SOLD, which is units sold in the year times the price paid to acquire or make the unit sold.
4. INCOME BEFORE TAXES, which equals sales less total costs.
5. INCOME TAX EXPENSE, which is zero if there are losses; otherwise, it is the income before taxes times the tax rate. (INCOME TAX EXPENSE is sometimes called merely INCOME TAXES.)
6. NET INCOME AFTER TAXES, which equals income before taxes less income tax expense.
7. END-OF-THE-YEAR CASH ON HAND is beginning-of-the-year cash on hand plus net income. (In the real world, cash flow estimates must account for changes in receivables and payables. In this case, assume that sales are collected immediately—that is, there are no receivables or bad debts. Assume also that suppliers are paid immediately—that is, that there are no payables.)

Construction of the Spreadsheet Model

Next, let's work through the following three steps to build your spreadsheet model:

1. Make a "skeleton" of the spreadsheet, and call it **TUTC.xls**.
2. Fill in the "easy" cell formulas.
3. Enter the "hard" spreadsheet formulas.

Make a Skeleton

Your first step is to set up a skeleton worksheet. This should have headings, text string labels, and constants—but no formulas.

To set up the skeleton, you must get a grip on the problem, *conceptually*. The best way to do that is to work *backward* from what the "body" of the spreadsheet will look like. Here, the body is the **INCOME STATEMENT AND CASH FLOW STATEMENT** section. Set that up, in your mind or on paper, then do the following:

- Decide what amounts should be in the **CALCULATIONS** section. In this tutorial's model, SALES (revenue) will be NUMBER OF UNITS SOLD IN A DAY times SELLING PRICE PER UNIT, in the income statement. You will calculate the intermediate amounts (NUMBER OF UNITS SOLD IN A YEAR and SELLING PRICE PER UNIT) in the **CALCULATIONS** section.
- Set up the **SUMMARY OF KEY RESULTS** section by deciding what *outputs* are needed to solve the problem. The **INPUTS** section should be reserved for amounts that can change—the controlling variables—which are the ECONOMIC OUTLOOK and the PURCHASE-PRICE OUTLOOK.
- Use the **CONSTANTS** section for values that you will need to use, but that are not in doubt, that is, you will not have to input them or calculate them. Here, the TAX RATE is a good example of such a value.

AT THE KEYBOARD

Type in the Excel skeleton shown in Figures C-1 and C-2, discussed in the fourth section of this tutorial.

A designation of "NA" means that a cell will not be used in any formula in the worksheet. The 2006 values are needed only for certain calculations, so for the most part, the 2006 column's cells just show "NA." (Recall that the forecast is for 2007 and 2008.) Also be aware that you can "break" a text string in a cell by pressing the keys Alt and Enter at the same time at the break point. This makes the cell "taller." Formatting of cells to show centered data and creation of borders is discussed at the end of this tutorial.

Fill in the "Easy" Formulas

The next step is to fill in the "easy" formulas. The cells affected (and what you should enter) are discussed next.

To prepare, you should format the cells in the **SUMMARY OF KEY RESULTS** section for no decimals. (Formatting for numerical precision is discussed at the end of this tutorial.) The **SUMMARY OF KEY RESULTS** section just "echoes" results shown in other places. The idea is that C28 holds the NET INCOME AFTER TAXES. You want to echo that amount in C12. So, the formula in C12 is =C28. Translation: "Copy what is in C28 into C12." It's that simple.

NOTE

With the insertion point in C12, the contents of that cell—in this case the formula =C28—shows in the editing window, which is above the lettered column indicators, as shown in Figure C-3.

C12 ▾ fx =C28

	A	B	C	D
11	SUMMARY OF KEY RESULTS	2006	2007	2008
12	NET INCOME AFTER TAXES	NA	0	
13	END-OF-THE-YEAR CASH ON HAND	NA		

Figure C-3 Echo 2007 NET INCOME AFTER TAXES

At this point, C28 is empty (and thus has a zero value), but that does not prevent you from copying. So, copy cell C12's formula to the right, to cell D12. (The copy operation does not actually "copy.") Copying puts =D28 into D12, which is what you want. (Year 2008's NET INCOME AFTER TAXES is in D28.)

To perform the Copy operation, use the following steps:

1. Select (click in) the cell (or range of cells) that you want to copy.
2. Choose **Edit—Copy**.
3. Select the cell (or cell range) to be copied to by clicking (and then dragging if the range has more than one cell).
4. Press the **Enter** key.

END-OF-THE-YEAR CASH ON HAND for 2006 cash is in cell C30. Echo the cash results in cell C30 to cell C13. (Put =C30 in cell C13, as shown in Figure C-4.) Copy the formula from C13 to D13.

C13 ▾ fx =C30

	A	B	C	D
11	SUMMARY OF KEY RESULTS	2006	2007	2008
12	NET INCOME AFTER TAXES	NA	0	0
13	END-OF-THE-YEAR CASH ON HAND	NA	0	

Figure C-4 Echo 2007 END-OF-THE-YEAR CASH ON HAND

At this point, the **CALCULATIONS** section formulas will not be entered because they are not all "easy" formulas. Move on to the easier formulas in the **INCOME STATEMENT AND CASH FLOW STATEMENT** section, as if the calculations were already done. Again, the empty **CALCULATIONS** section cells in this section do not stop you from entering formulas. You should now format the cells in the **INCOME STATEMENT AND CASH FLOW STATEMENT** section for zero decimals.

BEGINNING-OF-THE-YEAR CASH ON HAND is the cash on hand at the end of the *prior* year. In C22 for the year 2007, type =B30. See the "skeleton" you just entered, as shown in Figure C-5. Cell B30 has the END-OF-THE-YEAR CASH ON HAND for 2006.

	A	B	C	D
21	**INCOME STATEMENT AND CASH FLOW STATEMENT**	**2006**	**2007**	**2008**
22	BEGINNING-OF-THE-YEAR CASH ON HAND	NA	10000	
23				
24	SALES (REVENUE)	NA		
25	COST OF GOODS SOLD	NA		
26	INCOME BEFORE TAXES	NA		
27	INCOME TAX EXPENSE	NA		
28	NET INCOME AFTER TAXES	NA		
29				
30	END-OF-THE-YEAR CASH ON HAND (BEGINNING-OF-THE-YEAR CASH, PLUS NET INCOME AFTER TAXES)	10000		

C22 fx =B30

Figure C-5 Echo of END-OF-THE-YEAR CASH ON HAND for 2006 to BEGINNING-OF-THE-YEAR CASH ON HAND for 2007

Copy the formula in cell C22 to the right. SALES (REVENUE) is just NUMBER OF UNITS SOLD IN A YEAR times SELLING PRICE PER UNIT. In cell C24, enter =C17*C19, as shown in Figure C-6.

C24 fx =C17*C19

	A	B	C	D
15	**CALCULATIONS**	**2006**	**2007**	**2008**
16	NUMBER OF UNITS SOLD IN A DAY	1000		
17	SELLING PRICE PER UNIT	7.00		
18	COST OF GOODS SOLD PER UNIT	3.00		
19	NUMBER OF UNITS SOLD IN A YEAR	NA		
20				
21	**INCOME STATEMENT AND CASH FLOW STATEMENT**	**2006**	**2007**	**2008**
22	BEGINNING-OF-THE-YEAR CASH ON HAND	NA	10000	0
23				
24	SALES (REVENUE)	NA	0	
25	COST OF GOODS SOLD	NA		

Figure C-6 Enter the formula to compute 2007 SALES

The formula C17*C19 multiplies units sold for the year times the unit selling price. (Cells C17 and C19 are empty now, which is why SALES shows as zero after the formula is entered.) Copy the formula to the right, to D24.

COST OF GOODS SOLD is handled similarly. In C25, type =C18*C19. This equals NUMBER OF UNITS SOLD IN A YEAR times COST OF GOODS SOLD PER UNIT. Copy to the right.

In cell C26, the formula for INCOME BEFORE TAXES is =C24–C25. Enter the formula. Copy to the right.

In the United States, income taxes are only paid on positive income before taxes. In cell C27, shown in Figure C-7, the INCOME TAX EXPENSE is zero if the INCOME BEFORE TAXES is zero or less; else, INCOME TAX EXPENSE equals the TAX RATE times the INCOME BEFORE TAXES. The TAX RATE is a constant (in C4). An =IF() statement is needed to express this logic:

IF(INCOME BEFORE TAXES is <= 0, put zero tax in C27,
else in C27 put a number equal to multiplying the
TAX RATE times the INCOME BEFORE TAXES)

C26 stands for the concept INCOME BEFORE TAXES, and C4 stands for the concept TAX RATE. So, in Excel, substitute those cell addresses:

=IF(C26 <= 0, 0, C4 * C26)

Copy the income tax expense formula to the right.

In cell C28, NET INCOME AFTER TAXES is just INCOME BEFORE TAXES less INCOME TAX EXPENSE: =C26-C27. Enter and copy to the right.

The END-OF-THE-YEAR CASH ON HAND is BEGINNING-OF-THE-YEAR CASH ON HAND plus NET INCOME AFTER TAXES. In cell C30, enter =C22+C28. The **INCOME STATEMENT AND CASH FLOW STATEMENT** section at that point is shown in Figure C-7. Then, copy the formula to the right.

C30 | fx =C22+C28

	A	B	C	D
21	**INCOME STATEMENT AND CASH FLOW STATEMENT**	**2006**	**2007**	**2008**
22	BEGINNING-OF-THE-YEAR CASH ON HAND	NA	10000	10000
23				
24	SALES (REVENUE)	NA	0	0
25	COST OF GOODS SOLD	NA	0	0
26	INCOME BEFORE TAXES	NA	0	0
27	INCOME TAX EXPENSE	NA	0	0
28	NET INCOME AFTER TAXES	NA	0	0
29				
30	END-OF-THE-YEAR CASH ON HAND (BEGINNING-OF-THE-YEAR CASH, PLUS NET INCOME AFTER TAXES)	10000	10000	

Figure C-7 Status of INCOME STATEMENT AND CASH FLOW STATEMENT

Put in the "Hard" Formulas

The next step is to finish the spreadsheet by filling in the "hard" formulas.

AT THE KEYBOARD

First, in C8 enter an "O" (no quotation marks) for OPTIMISTIC, and in C9 enter "U" (no quotation marks) for UP. There is nothing magic about these particular values—they just give the worksheet formulas some input to process. Recall that the inputs will cover both 2007 and 2008. Make sure "NA" is in D8 and D9, just to remind yourself that these cells will not be used for input or by other worksheet formulas. Your **INPUTS** section should look like the one shown in Figure C-8.

	A	B	C	D
7	INPUTS	2006	2007	2008
8	ECONOMIC OUTLOOK (O = OPTIMISTIC; P = PESSIMISTIC)	NA	O	NA
9	PURCHASE-PRICE OUTLOOK (U = UP; D = DOWN)	NA	U	NA

Figure C-8 Entering two input values

Recall that cell addresses in the **CALCULATIONS** section are already referred to in formulas in the **INCOME STATEMENT AND CASH FLOW STATEMENT** section. The next step is to enter formulas for these calculations. Before doing that, format NUMBER OF UNITS SOLD IN A DAY and NUMBER OF UNITS SOLD IN A YEAR for zero decimals, and format SELLING PRICE PER UNIT and COST OF GOODS SOLD PER UNIT for two decimals.

The easiest formula in the **CALCULATIONS** section is the NUMBER OF UNITS SOLD IN A YEAR, which is the calculated NUMBER OF UNITS SOLD IN A DAY (in C16) times the NUMBER OF BUSINESS DAYS (in C5). In C19, enter =C5*C16, as shown in Figure C-9.

C19 fx =C5*C16

	A	B	C	D
1	TUTORIAL EXERCISE			
2				
3	CONSTANTS	2006	2007	2008
4	TAX RATE	NA	0.33	0.35
5	NUMBER OF BUSINESS DAYS	NA	300	300
6				
7	INPUTS	2006	2007	2008
8	ECONOMIC OUTLOOK (O = OPTIMISTIC; P = PESSIMISTIC)	NA	O	NA
9	PURCHASE-PRICE OUTLOOK (U = UP; D = DOWN)	NA	U	NA
10				
11	SUMMARY OF KEY RESULTS	2006	2007	2008
12	NET INCOME AFTER TAXES	NA	0	0
13	END-OF-THE-YEAR CASH ON HAND	NA	10000	10000
14				
15	CALCULATIONS	2006	2007	2008
16	NUMBER OF UNITS SOLD IN A DAY	1000		
17	SELLING PRICE PER UNIT	7.00		
18	COST OF GOODS SOLD PER UNIT	3.00		
19	NUMBER OF UNITS SOLD IN A YEAR	NA	0	

Figure C-9 Entering the formula to compute 2007 NUMBER OF UNITS SOLD IN A YEAR

Copy the formula to cell D19, for year 2008.

Assume that if the ECONOMIC OUTLOOK is OPTIMISTIC, the 2007 NUMBER OF UNITS SOLD IN A DAY will be 6% more than that in 2006; in 2008, they will be 6% more than that in 2007. Also assume that if the ECONOMIC OUTLOOK is PESSIMISTIC, the NUMBER OF UNITS SOLD IN A DAY in 2007 will be 1% less than those sold in 2006; in

2008, they will be 1% less than those sold in 2007. An =IF() statement is needed in C16 to express this idea:

IF(economy variable = OPTIMISTIC,
then NUMBER OF UNITS SOLD IN A DAY will go UP 6%,
else NUMBER OF UNITS SOLD IN A DAY will go DOWN 1%)

Substituting cell addresses:

=IF($C$8 = "O", 1.06 * B16, .99 * B16)

In Excel, quotation marks denote labels. The input is a one-letter label. So, the quotation marks around the **'O'** are needed. You should also note that multiplying by 1.06 results in a 6% increase, whereas multiplying by .99 results in a 1% decrease.

Enter the entire =IF formula into cell C16, as shown in Figure C-10. Absolute addressing is needed ($C$8) because the address is in a formula that gets copied, *and* you do not want this cell reference to change (to D8, which has the value "NA") when you copy the formula to the right. Absolute addressing maintains the $C$8 reference when the formula is copied. Copy the formula in C16 to D16 for 2008.

C16 =IF($C$8="O",1.06*B16,0.99*B16)

	A	B	C	D
15	CALCULATIONS	2006	2007	2008
16	NUMBER OF UNITS SOLD IN A DAY	1000	1060	
17	SELLING PRICE PER UNIT	7.00		
18	COST OF GOODS SOLD PER UNIT	3.00		

Figure C-10 Entering the formula to compute 2007 NUMBER OF UNITS SOLD IN A DAY

The SELLING PRICE PER UNIT is also a function of the ECONOMIC OUTLOOK. The two-part rule is (assume) as follows:

- If the ECONOMIC OUTLOOK is OPTIMISTIC, the SELLING PRICE PER UNIT in 2007 will be 1.07 times that of 2006; in 2008 it will be 1.07 times that of 2007.
- On the other hand, if the ECONOMIC OUTLOOK is PESSIMISTIC, the SELLING PRICE PER UNIT in 2007 and 2008 will equal the per-unit price in 2006 (that is, the price will not change).

Test your understanding of the selling price calculation by figuring out what the formula should be for cell C17. Enter it and copy to the right. You will need to use absolute addressing. (Can you see why?)

The COST OF GOODS SOLD PER UNIT is a function of the PURCHASE-PRICE OUTLOOK. The two-part rule is (assume) as follows:

- If the PURCHASE-PRICE OUTLOOK is UP ("U"), COST OF GOODS SOLD PER UNIT in 2007 will be 1.25 times that of year 2006; in 2008, it will be 1.25 times that of 2007.
- On the other hand, if the PURCHASE-PRICE OUTLOOK is DOWN ("D"), the multiplier in each year will be 1.01.

Again, to test your understanding, figure out what the formula should be in cell C18. Enter it and copy to the right. You will need to use absolute addressing.

Your selling price and cost of goods sold formulas, given OPTIMISTIC and UP input values, should yield the calculated values shown in Figure C-11.

	A	B	C	D
15	CALCULATIONS	2006	2007	2008
16	NUMBER OF UNITS SOLD IN A DAY	1000	1060	1124
17	SELLING PRICE PER UNIT	7.00	7.49	8.01
18	COST OF GOODS SOLD PER UNIT	3.00	3.75	4.69
19	NUMBER OF UNITS SOLD IN A YEAR	NA	318000	337080

Figure C-11 Calculated values given OPTIMISTIC and UP input values

Assume that you change the input values to PESSIMISTIC and DOWN. Your formulas should yield the calculated values shown in Figure C-12.

	A	B	C	D
15	CALCULATIONS	2006	2007	2008
16	NUMBER OF UNITS SOLD IN A DAY	1000	990	980
17	SELLING PRICE PER UNIT	7.00	7.00	7.00
18	COST OF GOODS SOLD PER UNIT	3.00	3.03	3.06
19	NUMBER OF UNITS SOLD IN A YEAR	NA	297000	294030

Figure C-12 Calculated values given PESSIMISTIC and DOWN input values

That completes the body of your spreadsheet! The values in the **CALCULATIONS** section ripple through the **INCOME STATEMENT AND CASH FLOW STATEMENT** section because the income statement formulas reference the calculations. Assuming inputs of OPTIMISTIC and UP, the income and cash flow numbers should now look like those in Figure C-13.

	A	B	C	D
21	INCOME STATEMENT AND CASH FLOW STATEMENT	2006	2007	2008
22	BEGINNING-OF-THE-YEAR CASH ON HAND	NA	10000	806844
23				
24	SALES (REVENUE)	NA	2381820	2701460
25	COST OF GOODS SOLD	NA	1192500	1580063
26	INCOME BEFORE TAXES	NA	1189320	1121398
27	INCOME TAX EXPENSE	NA	392476	392489
28	NET INCOME AFTER TAXES	NA	796844	728909
29				
30	END-OF-THE-YEAR CASH ON HAND (BEGINNING-OF-THE-YEAR CASH, PLUS NET INCOME AFTER TAXES)	10000	806844	1535753

Figure C-13 Completed INCOME STATEMENT AND CASH FLOW STATEMENT section

Scenario Manager

You are now ready to use the Scenario Manager to capture inputs and results as you play "what if" with the spreadsheet.

Note that there are four possible combinations of input values: O-U (Optimistic-Up), O-D (Optimistic-Down), P-U (Pessimistic-Up), and P-D (Pessimistic-Down). Financial results for each combination will be different. Each combination of input values can be referred to as a "scenario." Excel's Scenario Manager records the results of each combination of input values as a separate scenario and then shows a summary of all scenarios in a separate worksheet. These summary worksheet values can be used as a raw table of numbers, which could be printed or copied into a Word document. The table of data could then be the basis for an Excel chart, which could also be printed or put into a memorandum.

You have four possible scenarios for the economy and the purchase price of goods sold: Optimistic-Up, Optimistic-Down, Pessimistic-Up, and Pessimistic-Down. The four input-value sets produce different financial results. When you use the Scenario Manager, define the four scenarios, then have Excel (1) sequentially run the input values "behind the scenes," and then (2) put the results for each input scenario in a summary sheet.

When you define a scenario, you give the scenario a name and identify the input cells and input values. You do this for each scenario. Then you identify the output cells, so Excel can capture the outputs in a Summary Sheet.

At the Keyboard

To start, select Tools—Scenarios. This leads you to a Scenario Manager window. Initially, there are no scenarios defined, and Excel tells you that, as you can see in Figure C-14.

Scenario Manager

No Scenarios defined. Choose Add to add scenarios.

Show

Close

Add...

Figure C-14 Initial Scenario Manager window

With this window, you are able to add scenarios, delete them, or change (edit) them. Toward the end of the process, you are also able to create the summary sheet.

NOTE

When working with this window and its successors, do not hit the Enter key to navigate. Use mouse clicks to move from one step to the next.

To continue with defining a scenario: Click the Add button. In the resulting Add Scenario window, give the first scenario a name: OPT-UP. Then type in the input cells in the Changing cells window, here, C8:C9. (*Note*: C8 and C9 are contiguous input cells. Non-contiguous input cell ranges can be separated by a comma.) Excel may add dollar signs to the cell address—do not be concerned about this. The window should look like the one shown in Figure C-15.

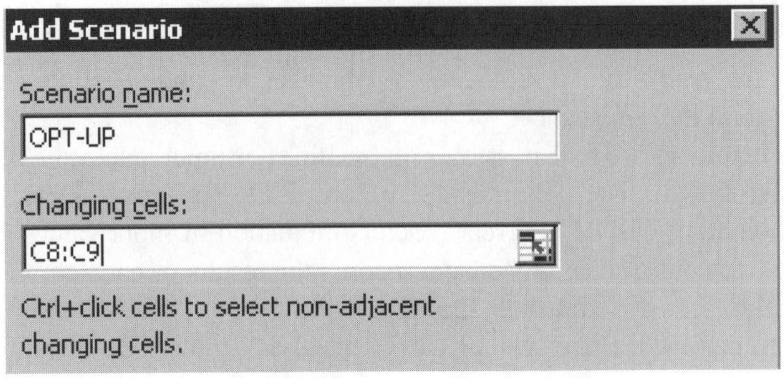

Figure C-15 Entering OPT-UP as a scenario

Now click OK. This moves you to the Scenario Values window. Here you indicate what the INPUT *values* will be for the scenario. The values in the cells *currently in* the spreadsheet will be displayed. They might—or might not—be correct for the scenario you are defining. For the OPT-UP scenario, an O and a U would need to be entered, if not the current values. Enter those values if need be, as shown in Figure C-16.

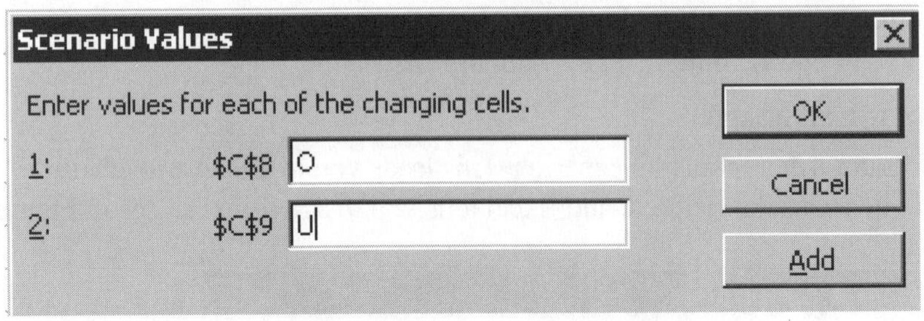

Figure C-16 Entering OPT-UP scenario input values

Now click OK. This takes you back to the Scenario Manager window. You are now able to enter the other three scenarios, following the same steps. Do that now! Enter the OPT-DOWN, PESS-UP, and PESS-DOWN scenarios, plus related input values. After all that, you should see that the names and the changing cells for the four scenarios have been entered, as in Figure C-17.

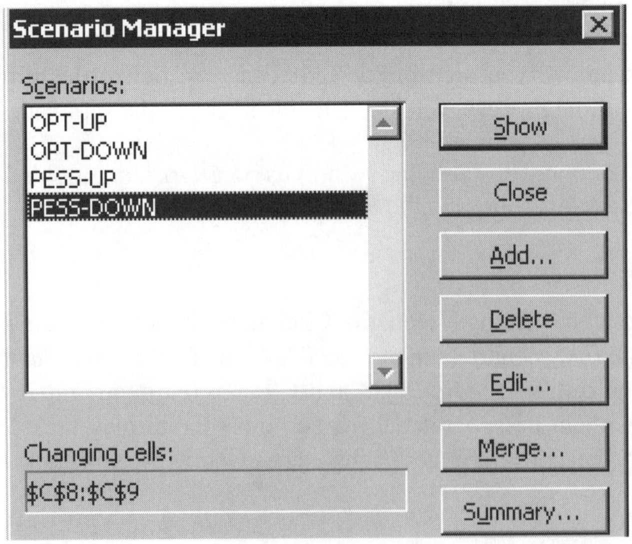

Figure C-17 Scenario Manager window with all scenarios entered

You are now able to create a summary sheet that shows the results of running the four scenarios. Click the Summary button. You'll get the Scenario Summary window. You must tell Excel what the output cell addresses are—these will be the same for all four scenarios. (The output *values* change in those output cells as input values are changed, but the addresses of the output cells do not change.)

Assume that you are primarily interested in the results that have accrued at the end of the two-year period. These are your two 2008 **SUMMARY OF KEY RESULTS** section cells for NET INCOME AFTER TAXES and END-OF-THE-YEAR CASH ON HAND (D12 and D13). Type these addresses into the window's input area, as shown in Figure C-18. (*Note*: If result cells are non-contiguous, the address ranges can be entered, separated by a comma.)

Scenario Summary

Report type

(•) Scenario summary

( ) Scenario PivotTable report

Result cells:

D12:D13

[OK] [Cancel]

Figure C-18 Entering Result cells addresses in Scenario Summary window

Then click OK. Excel runs each set of inputs and collects results as it goes. (You do not see this happening on the screen.) Excel makes a *new* sheet, titled the Scenario Summary (denoted by the sheet's lower tab), and takes you there, as shown in Figure C-19.

Scenario Summary	Current Values:	OPT-UP	OPT-DOWN	PESS-UP	PESS-DOWN
Changing Cells:					
$C$8	O	O	O	P	O
$C$9	U	U	D	U	U
Result Cells:					
$D$12	728909	728909	1085431	441964	728909
$D$13	1535753	1535753	2045679	1098681	1535753

Notes: Current Values column represents values of changing cells at time Scenario Summary Report was created. Changing cells for each scenario are highlighted in gray.

Figure C-19 Scenario Summary sheet created by Scenario Manager

One somewhat annoying visual element is that the Current Values in the spreadsheet itself are given an output column. This duplicates one of the four defined scenarios. You can delete that extra column by (1) clicking on its column designator letter (here, column D), and (2) clicking Edit—Delete.

Do *not* select Edit—Delete *Sheet*!

Tutorial C

Another annoyance is that Column A goes unused. You can click and delete it in the same way to move everything over to the left. This should make columns of data easier to see on the screen, without scrolling. You can also (1) edit cell values to make the results more clear, (2) enter words for the cell addresses, (3) use Alt—Enter to break long headings, if need be, (4) center values using the Format—Cells—Alignment tab menu option, and (5) show data in Currency format, Using the Format—Cells—Number tab menu option.

When you're done, your summary sheet should resemble the one shown in Figure C-20.

	A	B	C	D	E	F
1	**Scenario Summary**					
2			OPT-UP	OPT-DOWN	PESS-UP	PESS-DOWN
4	**Changing Cells:**					
5	**ECONOMIC OUTLOOK**	**$C$8**	O	O	P	P
6	**PURCHASE PRICE OUTLOOK**	**$C$9**	U	D	U	D
7	**Result Cells (2008):**					
8	**NET INCOME AFTER TAXES**	**$D$12**	$728,909	$1,085,431	$441,964	$752,953
9	**END-OF-THE-YEAR CASH ON HAND**	**$D$13**	$1,535,753	$2,045,679	$1,098,681	$1,552,944

Figure C-20 Scenario Summary sheet after formatting

Note that column C shows the OPTIMISTIC-UP case. NET INCOME AFTER TAXES in that scenario is $728,909, and END-OF-THE-YEAR CASH ON HAND is $1,535,753. Columns D, E, and F show the other scenario results.

Here is an important postscript to this exercise: DSS spreadsheets are used to guide decision-making. This means that the spreadsheet's results must be interpreted in some way. Let's practice with the results shown in Figure C-20. With that data, what combination of year 2008 NET INCOME AFTER TAXES and END-OF-THE-YEAR CASH ON HAND would be best? Clearly, OPTOMISTIC-DOWN (O-D) is the best result, right? It yields the highest income and highest cash. What is the worst combination? PESSIMISTIC-UP (P-U), right? It yields the lowest income and lowest cash.

Results are not always this easy to interpret, but the analytical method is the same. You have a complex situation that you cannot understand very well without software assistance. You build a model of the situation in the spreadsheet, enter the inputs, collect the results, and then interpret the results to guide decision-making.

Summary Sheets

When you do Scenario Manager spreadsheet case studies, you'll need to manipulate Summary Sheets and their data. Let's look at some of these operations.

Rerunning the Scenario Manager

The Scenario Summary sheet does not update itself if the spreadsheet formulas or inputs change. To see an updated Scenario Summary sheet, the user must rerun the Scenario Manager. To rerun the Scenario Manager, click the Summary button in the Scenario Manager dialog box and then click the OK button. This makes another summary sheet. It does not overwrite a prior one.

Deleting Unwanted Scenario Manager Summary Sheets

Suppose that you want to delete a Summary sheet. With the Summary sheet on the screen, select Edit—Delete Sheet. You will be asked if you really mean it. If so, click to remove, or else cancel out.

Charting Summary Sheet Data

The Summary sheet results can be conveniently charted using the Chart Wizard. Charting Excel data is discussed in Tutorial E.

Copying Summary Sheet Data to the Clipboard

You may want to put the summary sheet data into the Clipboard to use later in a Word document. To do that, use the following steps:

1. Highlight the data range.
2. Use **Edit—Copy** to put the graphic into the Clipboard.
3. Assuming that you want to exit Excel, select **File—Save**, **File—Close**, **File—Exit** Excel. (You may be asked whether you want to leave your data in the Clipboard—you do want to.)
4. Open your Word document.
5. Put your cursor where you want the upper-left part of the graphic to be positioned.
6. Select **Edit—Paste**.

PRACTICE USING SCENARIO MANAGER

Suppose that you have an uncle who works for a large company. He has a good job and makes a decent salary ($80,000 a year, currently). He can retire from his company in 2013, when he will be 65. He would start drawing his pension then.

However, the company has an "early out" plan. Under this plan, the company asks employees to quit (called "preretirement"). The company then pays those employees a bonus in the year they quit and each year thereafter, up to the official retirement date, which is through the year 2012 for your uncle. Then, employees start to receive their actual pension—in your uncle's case, in 2013. This "early out" program would let your uncle leave the company before 2013. Until then, he could find a part-time hourly-wage job to make ends meet and then leave the workforce entirely in 2013.

The opportunity to leave early is open through 2012. Your uncle could stay with the company in 2007, then preretire any time in the years 2008 to 2012, getting the "early out" bonuses in those years. Of course, if he retires in 2008, he would lose the 2007 bonus, and so on, all the way through 2012.

Another factor in your uncle's thinking is whether to continue belonging to his country club. He likes the club, but it is a real cash drain. The "early out" decision can be looked at each year, but the country club membership decision must be made now—if he does not withdraw in 2007, then he says he will stay a member (and incur costs) through 2012.

Your uncle has called you in to make a Scenario Manager spreadsheet model of his situation. Your spreadsheet would let him play "what if" with the preretirement and country club possibilities to see projected 2007-2012 personal finance results. He wants to know what "cash on hand" will be available for each year in the period with each scenario.

Complete the spreadsheet for your uncle. Your **SUMMARY OF KEY RESULTS**, **CALCULATIONS**, and **INCOME STATEMENT AND CASH FLOW STATEMENT** section cells must show values *by cell formula*. That is, in those areas, do not hard-code amounts. In any of your formulas, do not use the address of a cell if its contents are "NA." Set up your spreadsheet skeleton as shown in the figures that follow. Name your spreadsheet **UNCLE.xls**.

CONSTANTS Section

Your spreadsheet should have the constants shown in Figure C-21. An explanation of line items follows the figure.

	A	B	C	D	E	F	G	H
1	**YOUR UNCLE'S EARLY RETIREMENT DECISION**							
2	CONSTANTS	2006	2007	2008	2009	2010	2011	2012
3	CURRENT SALARY	80000	NA	NA	NA	NA	NA	NA
4	SALARY INCREASE FACTOR	NA	0.03	0.03	0.02	0.02	0.01	0.01
5	PART TIME WAGES EXPECTED	NA	10000	10200	10500	10800	11400	12000
6	BUY OUT AMOUNT	NA	30000	25000	20000	15000	5000	0
7	COST OF LIVING (NOT RETIRED)	NA	41000	42000	43000	44000	45000	46000
8	COUNTRY CLUB DUES	NA	12000	13000	14000	15000	16000	17000

Figure C-21 CONSTANTS section values

- SALARY INCREASE FACTOR: Your uncle's salary at the end of 2006 will be $80,000. As you can see, raises are expected in each year—for example, a 3% raise is expected in 2007. If he does not retire, he would get his salary and the small raise in a year.
- PART-TIME WAGES EXPECTED: Your uncle has estimated his part-time wages if he were retired and working part time in the 2007–2012 period.
- BUY OUT AMOUNT: The company's preretirement "buy out" plan amounts are shown. If your uncle retires in 2007, he gets $30,000, $25,000, $20,000, $15,000, $5,000, and zero in the years 2007 to 2012, respectively. If he leaves in 2008, he gives up the $30,000 2007 payment, but would get $25,000, $20,000, $15,000, $5,000, and zero in the years 2008 to 2012, respectively.
- COST OF LIVING: Your uncle has estimated how much cash he needs to meet his living expenses, assuming that he continues to work for the company. His cost of living would be $41,000 in 2007, increasing each year thereafter.
- COUNTRY CLUB DUES: Country club dues are $12,000 for 2007. They increase each year thereafter.

INPUTS Section

Your spreadsheet should have the inputs shown in Figure C-22. An explanation of line items follows the figure.

	A	B	C	D	E	F	G	H
10	INPUTS	2006	2007	2008	2009	2010	2011	2012
11	RETIRED [R] or WORKING [W]	NA						
12	STAY IN CLUB? [Y] OR [N]	NA		NA	NA	NA	NA	NA

Figure C-22 INPUTS section

- RETIRED OR WORKING: Enter an "R" if your uncle retires in a year, or a "W" if he is still working. If he is working through 2012, the pattern **WWWWWW** should be entered. If his retirement is in 2007, the pattern **RRRRRR** should be entered. If he works for three years and then retires in 2010, the pattern **WWWRRR** should be entered.
- STAY IN CLUB?: If your uncle stays in the club in 2007–2012, a "**Y**" should be entered. If your uncle is leaving the club in 2007, an "**N**" should be entered. The decision applies to all years.

SUMMARY OF KEY RESULTS Section

Your spreadsheet should show the results in Figure C-23.

	A	B	C	D	E	F	G	H
14	SUMMARY OF KEY RESULTS	2006	2007	2008	2009	2010	2011	2012
15	END-OF-THE-YEAR CASH ON HAND	NA						

Figure C-23 SUMMARY OF KEY RESULTS section

Each year's END-OF-THE-YEAR CASH ON HAND value is echoed from cells in the spreadsheet body.

CALCULATIONS Section

Your spreadsheet should calculate, by formula, the values shown in Figure C-24. Calculated amounts are used later in the spreadsheet. An explanation of line items follows the figure.

	A	B	C	D	E	F	G	H
17	CALCULATIONS	2006	2007	2008	2009	2010	2011	2012
18	TAX RATE	NA						
19	COST OF LIVING	NA						
20	YEARLY SALARY OR WAGES	80000						
21	COUNTRY CLUB DUES PAID	NA						

Figure C-24 CALCULATIONS section

- TAX RATE: Your uncle's tax rate depends on whether he is retired. Retired people have lower overall tax rates. If he is retired in a year, your uncle's rate is expected to be 15% of income before taxes. In a year in which he works full time, the rate will be 30%.
- COST OF LIVING: In any year that your uncle continues to work for the company, his cost of living is what is shown in COST OF LIVING (NOT RETIRED) in the **CONSTANTS** section in Figure C-21. But if he chooses to retire, his cost of living is $15,000 less than the amount shown in the figure.
- YEARLY SALARY OR WAGES: If your uncle keeps working, his salary increases each year. The year-to-year percentage increases are shown in the **CONSTANTS** section. Thus, salary earned in 2007 would be more than that earned in 2006, salary earned in 2008 would be more than that earned in 2007, and so on. If your uncle retires in a certain year, he will make the part-time wages shown in the **CONSTANTS** section.
- COUNTRY CLUB DUES PAID: If your uncle leaves the club, the dues are zero each year; otherwise, the dues are as shown in the **CONSTANTS** section.

The INCOME STATEMENT AND CASH FLOW STATEMENT Section

This section begins with the cash on hand at the beginning of the year. This is followed by the income statement, concluding with the calculation of cash on hand at the end of the year. The format is shown in Figure C-25. An explanation of line items follows the figure.

	A	B	C	D	E	F	G	H
23	**INCOME STATEMENT AND CASH FLOW STATEMENT**	**2006**	**2007**	**2008**	**2009**	**2010**	**2011**	**2012**
24	BEGINNING-OF-THE-YEAR CASH ON HAND	**NA**						
25								
26	SALARY OR WAGES	**NA**						
27	BUY OUT INCOME	**NA**						
28	TOTAL CASH INFLOW	**NA**						
29	COUNTRY CLUB DUES PAID	**NA**						
30	COST OF LIVING	**NA**						
31	TOTAL COSTS	**NA**						
32	INCOME BEFORE TAXES	**NA**						
33	INCOME TAX EXPENSE	**NA**						
34	NET INCOME AFTER TAXES	**NA**						
35								
36	END-OF-THE-YEAR CASH ON HAND (BEGINNING-OF-THE-YEAR CASH, PLUS NET INCOME AFTER TAXES)	30000						

Figure C-25 INCOME STATEMENT AND CASH FLOW STATEMENT section

- BEGINNING-OF-THE-YEAR CASH ON HAND: This is the END-OF-THE-YEAR CASH ON HAND at the end of the prior year.
- SALARY OR WAGES: This is a yearly calculation, which can be echoed here.
- BUY OUT INCOME: This is the year's "buy out" amount, if your uncle is retired in the year.
- TOTAL CASH INFLOW: This is the sum of salary or part-time wages and "buy out" amounts.
- COUNTRY CLUB DUES PAID: This is a calculated amount.
- COST OF LIVING: This is a calculated amount.
- TOTAL COSTS: These outflows are the sum of the COST OF LIVING and COUNTRY CLUB DUES PAID.
- INCOME BEFORE TAXES: This amount is the TOTAL CASH INFLOW less TOTAL COSTS (outflows).
- INCOME TAX EXPENSE: This amount is zero if INCOME BEFORE TAXES is zero or less; otherwise, the calculated tax rate is applied to the INCOME BEFORE TAXES.
- NET INCOME AFTER TAXES: This is INCOME BEFORE TAXES, less TAX EXPENSE.
- END-OF-THE-YEAR CASH ON HAND: This is the BEGINNING-OF-THE-YEAR CASH plus the year's NET INCOME AFTER TAXES.

Scenario Manager Analysis

Set up the Scenario Manager and create a Scenario Summary sheet. Your uncle wants to look at the following four possibilities:

- Retire in 2007, staying in the club ("Loaf-In")
- Retire in 2007, leaving the club ("Loaf-Out")
- Work three more years, retire in 2010, staying in the club ("Delay-In")
- Work three more years, retire in 2010, leaving the club ("Delay-Out")

You can enter non-contiguous cell ranges as follows: C20..F20, C21, C22 (cell addresses are examples). The output cell should be the 2012 (only) END-OF-THE-YEAR CASH ON HAND cell.

Your uncle will choose the option that yields the highest 2012 END-OF-THE-YEAR CASH ON HAND. You must look at your Scenario Summary sheet to see which strategy yields the highest amount.

To check your work, you should attain the values shown in Figure C-26. (You can use the labels Excel gives you in the left-most column or change the labels, as was done in Figure C-26.)

	A	B	C	D	E	F
1	Scenario Summary					
2			LOAF-IN	LOAF-OUT	DELAY-IN	DELAY-OUT
4	Changing Cells:					
5	RETIRE OR WORK, 2007	$C$11	R	R	W	W
6	RETIRE OR WORK, 2008	$D$11	R	R	W	W
7	RETIRE OR WORK, 2009	$E$11	R	R	W	W
8	RETIRE OR WORK, 2010	$F$11	R	R	R	R
9	RETIRE OR WORK, 2011	$G$11	R	R	R	R
10	RETIRE OR WORK, 2012	$H$11	R	R	R	R
11	IN CLUB 2007-2012?	$C$12	Y	N	Y	N
12	Result Cells (2012):					
13	END-OF-THE-YEAR CASH ON HAND	$H$15	-$68,400	$15,195	$8,389	$83,689

Figure C-26 Scenario Summary

Review of Excel Basics

In this section, you'll begin by reviewing how to perform some basic operations. Then, you'll work through some further cash flow calculations. Reading and working through this section will help you to do the spreadsheet cases in this book.

Basic Operations

In this section, you'll review the following topics: formatting cells, showing Excel cell formulas, understanding circular references, using the And and the Or functions in IF statements, and using nested IF statements.

Formatting Cells

You may have noticed that some data in this tutorial's first spreadsheet was centered in the cells. Here is how to perform that operation:

1. Highlight the cell range to format.
2. Select the **Format** menu option.
3. Select **Cells—Alignment**.
4. Choose **Center** for both **Horizontal** and **Vertical**.
5. Select **OK**.

It is also possible to put a border around cells. This treatment might be desirable for highlighting **INPUTS** section cells. To perform this operation:

1. Select **Format—Cells—Border—Outline**.
2. Choose the outline **Style** you want.
3. Select **OK**.

You can format numerical values for Currency format by selecting the following:

Format—Cells—Number—Currency.

You can format numerical values for decimal places using this procedure:

1. Select **Format—Cells—Number** tab—**Number**.
2. Select the desired number of decimal places.

Showing Excel Cell Formulas

If you want to see Excel cell formulas, follow this procedure:

1. Press the **Ctrl** key and the "back-quote" key (**`**) at the same time. (The back-quote orients from Seattle to Miami—on most keyboards, it is next to the exclamation-point key and shares the key with the tilde diacritic mark.)
2. To restore, press the **Ctrl** key and (**`**) back-quote key again.

Understanding a Circular Reference

A formula has a circular reference if the *formula refers to the cell that the formula is in*. Excel cannot properly evaluate such a formula, because the value of the cell is not yet known—but to do that evaluation, the value in the cell must already be known! The reasoning is circular, hence the term "circular reference." Excel will point out circular references, if any exist, when you choose Open for a spreadsheet. Excel will also point out circular references as you insert them during development. Excel will be demonstrative about this by opening at least one Help window and by drawing arrows between cells involved in the offending formula. You can close the windows, but that will not fix the situation. You *must* fix the formula that has the circular reference if you want the spreadsheet to give you accurate results.

Here is an example. Suppose that the formula in cell C18 is =C18 – C17. Excel tries to evaluate the formula in order to put a value on the screen in cell C18. To do that, Excel must know the value in cell C18—but that is what it is trying to do in the first place. Can you see the circularity?

Using the "And" Function and the "Or" Function in =IF Statements

An =IF() statement has the following syntax:

=IF(test condition, result if test is True, result if test is False)

The test conditions in this tutorial's =IF statements tested only one cell's value. A test condition could test more than one cell's values.

Here is an example from this tutorial's first spreadsheet. In that example, selling price was a function of the economy. Assume, for the sake of illustration, that year 2007's selling price per unit depends on the economy *and* the purchase price outlook. If the economic outlook is optimistic *and* the company's purchase price outlook is down, then the selling price will be 1.10 times the prior year's price. Assume that in all other cases, the selling price will be 1.03 times the prior year's price. The first test requires two things to be true *at the same time*: C8 = "O" *AND* C9 = "D." So, the AND() function would be needed. The code in cell C17 would be as follows:

=IF(AND($C$8 = "O", $C$9 = "D"), 1.10 * B17, 1.03 * B17)

On the other hand, the test might be this: If the economic outlook is optimistic *or* the purchase price outlook is down, then the selling price will be 1.10 times the prior year's price. Assume that in all other cases, the selling price will be 1.03 times the prior year's price. The first test requires *only one of* two things to be true: C8 = "O" *or* C9 = "D". Thus, the OR() function would be needed. The code in cell C17 would be:

=IF(OR($C$8 = "O", $C$9 = "D"), 1.10 * B17, 1.03 * B17)

Using IF() Statements Inside IF() Statements

An =IF() statement has this syntax:

=IF(test condition, result if test is True, result if test is False)

In the examples shown thus far, only two courses of action were possible, so only one test has been needed in the =IF() statement. There can be more courses of action than two, however, and this requires that the "result if test is False" clause needs to show further testing. Let's look at an example.

Assume again that the 2007 selling price per unit depends on the economy and the purchase price outlook. Here is the logic: (1) If the economic outlook is optimistic *and* the purchase price outlook is down, then the selling price will be 1.10 times the prior year's price. (2) If the economic outlook is optimistic *and* the purchase price outlook is up, then the selling price will be 1.07 times the prior year's price. (3) In all other cases, the selling price will be 1.03 times the prior year's price. The code in cell C17 would be:

=IF(AND($C$8 = "O", $C$9 = "D"), 1.10 * B17,
IF(AND($C$8 = "O", $C$9 = "U"), 1.07 * B17, 1.03 * B17))

The first =IF() tests to see if the economic outlook is optimistic and the purchase price outlook is down. If not, further testing is needed to see whether the economic outlook is optimistic and the purchase price outlook is up, or whether some other situation prevails.

NOTE

Be sure to note the following:

- The line is broken in the previous example because the page is not wide enough, but in Excel, the formula would appear on one line.
- The embedded "IF" is not preceded by an equals sign.

Example: Borrowing and Repayment of Debt

The Scenario Manager cases require you to account for money that the company borrows or repays. Borrowing and repayment calculations are discussed next. At times you are asked to think about a question and fill in the answers. Correct responses are at the end of this section.

To do the Scenario Manager cases, you must assume two things about a company's borrowing and repayment of debt. First, assume that the company wants to have a certain minimum cash level at the end of a year (and thus to start the next year). Assume that a bank will provide a loan to reach the minimum cash level if year-end cash falls short of that level.

Here are some numerical examples to test your understanding. Assume that NCP stands for "net cash position" and equals beginning-of-the-year cash plus net income after taxes for the year. The NCP is the cash available at year end, before any borrowing or repayment. Compute the amounts to borrow in the three examples in Figure C-27.

Example	NCP	Minimum Cash Required	Amount to Borrow
1	50,000	10,000	?
2	8,000	10,000	?
3	–20,000	10,000	?

Figure C-27 Examples of borrowing

Tutorial C

Assume that a company would use its excess cash at year end to pay off as much debt as possible, without going below the minimum-cash threshold. "Excess cash" would be the NCP *less* the minimum cash required on hand—amounts over the minimum are available to repay any debt.

In the examples shown in Figure C-28, compute excess cash and then compute the amount to repay. You may also want to compute ending cash after repayments as well, to aid your understanding.

Example	NCP	Minimum Cash Required	Beginning-of-the-Year Debt	Repay	Ending Cash
1	12,000	10,000	4,000	?	?
2	12,000	10,000	10,000	?	?
3	20,000	10,000	10,000	?	?
4	20,000	10,000	0	?	?
5	60,000	10,000	40,000	?	?
6	–20,000	10,000	10,000	?	?

Figure C-28 Examples of repayment

In this section's Scenario Manager cases, your spreadsheet will need two bank financing sections beneath the **INCOME STATEMENT AND CASH FLOW STATEMENT** section:

1. The first section will calculate any needed borrowing or repayment at the year's end to compute year-end cash.
2. The second section will calculate the amount of debt owed at the end of the year, after borrowing or repayment of debt.

The first new section, in effect, extends the end-of-year cash calculation, which was shown in Figure C-13. Previously, the amount equaled cash at the beginning of the year plus the year's net income. Now, the calculation will include cash obtained by borrowing and cash repaid. Figure C-29 shows the structure of the calculation.

	A	B	C	D
30	NET CASH POSITION (NCP) BEFORE BORROWING AND REPAYMENT OF DEBT (BEGINNING-OF-THE-YEAR CASH PLUS NET INCOME AFTER TAXES	NA		
31	PLUS: BORROWING FROM BANK	NA		
32	LESS: REPAYMENT TO BANK	NA		
33	EQUALS: END-OF-THE-YEAR CASH ON HAND	10000		

Figure C-29 Calculation of END-OF-THE-YEAR CASH ON HAND

The heading in cell A30 was previously END-OF-THE-YEAR CASH ON HAND. But BORROWING increases cash and REPAYMENT OF DEBT decreases cash. So, END-OF-THE-YEAR CASH ON HAND is now computed two rows down (in C33 for year 2007, in the example). The value in row 30 must be a subtotal for the BEGINNING-OF-THE-YEAR CASH ON HAND plus the year's NET INCOME AFTER TAXES. We call that subtotal the NET CASH POSITION (NCP) BEFORE BORROWING AND REPAYMENT OF DEBT.

(*Note*: Previously, the formula in cell C22 for BEGINNING-OF-THE-YEAR CASH ON HAND was =B30. Now, that formula would be =B33. It would be copied to the right, as before, for the next year.)

That second new section would look like what is shown in Figure C-30.

	A	B	C	D
35	DEBT OWED	2006	2007	2008
36	BEGINNING-OF-THE-YEAR DEBT OWED	NA		
37	PLUS: BORROWING FROM BANK	NA		
38	LESS: REPAYMENT TO BANK	NA		
39	EQUALS: END-OF-THE-YEAR DEBT OWED	15000		

Figure C-30 DEBT OWED section

The second new section computes end-of-year debt and is called DEBT OWED. At the end of 2006, $15,000 was owed. END-OF-THE-YEAR DEBT OWED equals the BEGINNING-OF-THE-YEAR DEBT OWED, plus any new BORROWING FROM BANK (which increases debt owed), less any REPAYMENT TO BANK (which reduces it). So, in the example, the formula in cell C39 would be:

=C36 + C37 – C38

Assume that the amounts for BORROWING FROM BANK and REPAYMENT TO BANK are calculated in the first new section. Thus, the formula in cell C37 would be: =C31. The formula in cell C38 would be: =C32. (BEGINNING-OF-THE-YEAR DEBT OWED is equal to the debt owed at the end of the prior year, of course. The formula in cell C36 for BEGINNING-OF-THE-YEAR DEBT OWED would be an echoed formula. *Can you see what it would be*? It's an exercise for you to complete. *Hint*: The debt owed at the beginning of a year equals the debt owed at the end of the prior year.)

Now that you have seen how the borrowing and repayment data is shown, the logic of the borrowing and repayment formulas can be discussed.

Calculation of BORROWING FROM BANK

The logic of this in English is:

If (cash on hand before financing transactions is greater than the minimum cash required, then borrowing is not needed; otherwise, borrow enough to get to the minimum).

Or (a little more precisely):

If (NCP is greater than the minimum cash required, then BORROWING FROM BANK = 0; otherwise, borrow enough to get to the minimum).

Suppose that the desired minimum cash at year end is $10,000, and that value is a constant in your spreadsheet's cell C6. Assume the NCP is shown in your spreadsheet's cell C30. Our formula (getting closer to Excel) would be as follows:

IF(NCP > Minimum Cash, 0; otherwise, borrow enough to get to the minimum).

Tutorial C

You have cell addresses that stand for NCP (cell C30) and Minimum Cash (C6). To develop the formula for cell C31, substitute these cell addresses for NCP and Minimum Cash. The harder logic is that for the "otherwise" clause. At this point, you should look ahead to the borrowing answers at the end of this section, Figure C-31. In Example 2, $2,000 had to be borrowed. Which cell was subtracted from which other cell to calculate that amount? Substitute cell addresses in the Excel formula for the year 2007 borrowing formula in cell C31:

=IF(>=   , 0 ,   –   )

The Answer is at the end of this section, Figure C-33.

Calculation of REPAYMENT TO BANK

The logic of this in English is:

IF(beginning of year debt = 0, repay 0 because nothing is owed, but
    IF(NCP is less than the min, repay zero, because you must *borrow*, but
        IF(extra cash equals or exceeds the debt, repay the whole debt,
            ELSE (to stay above the min, repay only the extra cash))))

Look at the following formula. Assume the repayment will be in cell C32. Assume debt owed at the beginning of the year is in cell C36, and that minimum cash is in cell C6. Substitute cell addresses for concepts to complete the formula for year 2007 repayment. (Clauses are on different lines because of page width limitations.)

=IF(= 0, 0,
    IF(<=   , 0,
        IF((–   ) >=   ,
            (–   )))).

The answer is at the end of this section, in Figure C-34.

Answers to Questions About Borrowing and Repayment Calculations

Figures C-31 and C-32 give the answers to the questions posed about borrowing and repayment calculations.

Example	NCP	Minimum Cash Required	Borrow	Comments
1	50,000	10,000	Zero	NCP > Min.
2	8,000	10,000	2,000	Need 2000 to get to Min. 10,000 – 8,000
3	–20,000	10,000	30,000	Need 30000 to get to Min. 10,000 – (–20,000)

Figure C-31 Answers to examples of borrowing

Example	NCP	Minimum Cash Required	Beginning-of-the-Year Debt	Repay	Ending Cash
1	12,000	10,000	4,000	2,000	10,000
2	12,000	10,000	10,000	2,000	10,000
3	20,000	10,000	10,000	10,000	10,000
4	20,000	10,000	0	0	20,000
5	60,000	10,000	40,000	40,000	20,000
6	–20,000	10,000	10,000	NA	NA

Figure C-32 Answers to examples of repayment

Some notes about the repayment calculations shown in Figure C-32 follow.

- In Examples 1 and 2, only $2,000 is available for debt repayment (12,000 – 10,000) to avoid going below the minimum cash.
- In Example 3, cash available for repayment is $10,000 (20,000 – 10,000), so all beginning debt can be repaid, leaving the minimum cash.
- In Example 4, there is no debt owed, so no debt need be repaid.
- In Example 5, cash available for repayment is $50,000 (60,000 – 10,000), so all beginning debt can be repaid, leaving more than the minimum cash.
- In Example 6, no cash is available for repayment. The company must borrow.

Figures C-33 and C-34 show the calculations for borrowing and repayment of debt.

```
=IF( C30 >= C6, 0, C6 – C30)
```

Figure C-33 Calculation of borrowing

```
=IF( C36 = 0, 0, IF( C30 <= C6, 0, IF( (C30 – C6) >= C36, C36, ( C30 – C6) )))
```

Figure C-34 Calculation of repayment

Saving Files After Using Microsoft Excel

As you work, save periodically (**File—Save**). If you want to save to a disk, choose **Drive A:**. At the end of a session, save your work using this three-step procedure:

1. Save the file, using **File—Save**. If you want to save to a disk, choose **Drive A:**.
2. Use **File—Close** to tell Windows to close the file. If saving to a disk, make sure it is still in **Drive A:** when you close. If you violate this rule, you may lose your work!

CAUTION

3. Exit from Excel to Windows by selecting **File—Exit**. In theory, you may exit from Excel back to Windows after you have saved a file (short-cutting the File—Close step), but that is not a recommended shortcut.

Eastern Coffee's Labor Negotiation Budget

DECISION SUPPORT USING EXCEL

PREVIEW

Eastern Coffee Company is heading into crucial labor negotiations with its union. In this case, you will use Excel to model the financial implications of some pay rate and work rule proposals.

PREPARATION

- Review spreadsheet concepts discussed in class and/or in your textbook.
- Complete any exercises that your instructor assigns.
- Complete any part of Tutorial C that your instructor assigns, or refer to it as necessary.
- Review file-saving procedures for Windows programs. These are discussed in Tutorial C.
- Refer to Tutorial E as necessary.

BACKGROUND

The Eastern Coffee Company makes coffee pots. The pots typically have a glass body and a plastic top and handle. The pots are sold to restaurant chains and to other companies who sell them under their own brand name.

The company's labor force is unionized. All the employees belong to the UCWA—the United Coffee Workers of America. The company and the union are negotiating the next labor agreement. The labor agreement would cover the period 2007–2009.

Key Variables

The two key variables in the negotiation are: (1) the percentage of pay increase; and (2) changes to shop floor "work rules." The workers want a pay increase. The company wants more flexibility in assigning workers to jobs and in other operational matters. In prior years, the union has insisted on restrictive work rules. The company would like to be able to use idle coffee pot assemblers for certain non-assembly jobs. Currently, that sort of assignment cannot be made—an assembler can only put pots together, a janitor can only clean, and so on. If the company could get the union to relax work rules, efficiency would increase, and many costs would go down.

A third variable, one not open to negotiation, is the state of the economy. The state of the economy in years 2007–2009 is not known, but it will go up, remain steady, or go down.

Impact of Key Variables

Key variables are predicted to have specific impacts. For instance, giving workers a pay increase will increase direct labor costs. On the other hand, if the union cooperates on granting less-restrictive work rules, then other variable costs will go down; if the union will not cooperate, other variable costs will go up.

Management thinks they may have to "buy" work-rule cooperation. In other words, they may have to give workers a healthy pay increase to get work-rule cooperation. On the other hand, if management insists on a low pay increase, then the union is unlikely to agree to relax restrictive work rules. Conceivably, in a good economy, the company can have good financial results even if work rules are not relaxed.

The state of the economy will affect the following:

- The interest rate on debt the company owes to its bank
- The selling price of a coffee pot
- The number of pots sold

The company needs a spreadsheet that will show a budgeted income statement and cash flow, given assumptions about the economy, pay rate changes, and whether the union will agree to relax work rules.

You have been called in to make a projected 2007–2009 budget in Excel. Certain scenarios are under consideration, and your spreadsheet will model them in the Scenario Manager.

ASSIGNMENT 1 CREATING A SPREADSHEET FOR DECISION SUPPORT

In this assignment, you will produce a spreadsheet that models the business decision. Then, in Assignment 2, you will write a memorandum to the company's CEO that explains your recommended action. In addition, in Assignment 3, you will be asked to prepare an oral presentation of your analysis and recommendation.

First, you will create the spreadsheet model of the negotiation. The model is an analysis for the years 2007–2009. You will be given some hints on how each section should be set up before entering cell formulas. Your spreadsheet should have the sections that follow:

- CONSTANTS
- INPUTS
- SUMMARY OF KEY RESULTS
- CALCULATIONS
- INCOME STATEMENT AND CASH FLOW STATEMENT

A discussion of each section follows. *The spreadsheet skeleton is available to you, so you need not type in the skeleton if you do not wish to do so.* To access the spreadsheet, go to your Data files. Select Case 6, and then select POTS.xls.

CONSTANTS Section

Your spreadsheet should have the constants shown in Figure 6-1. An explanation of the line items follows the figure.

	A	B	C	D	E
1	EASTERN COFFEE CORPORATION BUDGET ANALYSIS				
2					
3	CONSTANTS	2006	2007	2008	2009
4	TAX RATE	NA	0.32	0.33	0.34
5	FIXED COSTS	NA	4000000	4100000	4200000
6	CASH NEEDED TO START NEXT YEAR	NA	4000000	4000000	4000000
7	RAW MATERIAL COST PER UNIT	NA	1.00	1.05	1.10

Figure 6-1 CONSTANTS section

- TAX RATE: The expected tax rates on income before taxes for 2007–2009 are shown.
- FIXED COSTS: Fixed costs are expected to increase in the next few years, as shown.
- CASH NEEDED TO START NEXT YEAR: The company's bankers say they need to start each year with a certain amount of cash, as shown.
- RAW MATERIAL COST PER UNIT: On average, raw materials will cost $1 per coffee pot in 2007, increasing 5 cents each year thereafter.

INPUTS Section

Your spreadsheet should have the inputs shown in Figure 6-2. An explanation of the line items follows the figure.

	A	B	C	D	E
9	INPUTS	2006	2007	2008	2009
10	EXPECTED ECONOMY (U = UP, S = STEADY, D = DOWN)	NA			
11	PAY INCREASE (.XX)		NA	NA	NA
12	WORK RULE AGREEMENT (YES / NO)		NA	NA	NA

Figure 6-2 INPUTS section

- EXPECTED ECONOMY: The spreadsheet user enters the expected state of the economy in each of the next three years. If the economy is expected to be strong, the user enters a *U* for *Up*. If the economy is expected to be steady, the user enters an *S* for *Steady*. If the economy is expected to worsen, the user enters a *D* for *Down*. The sequence *SSS* would mean a steady economy in all years 2007–2009. The sequence *DDU* would mean a weak economy in 2007–2008 but a strong economy in 2009.
- PAY INCREASE: The amount of the average pay increase for all three years is entered directly. For example, if workers will on average get an 8% pay increase, .08 would be entered. If a pay decrease of 2% was negotiated, –.02 would be entered.
- WORK RULE AGREEMENT (YES / NO): If the workers will agree to relaxing the work rules, the word *YES* would be entered. If not, the word *NO* would be entered.

Your instructor may tell you to apply Conditional Formatting to the input cells, so out-of-bounds values are highlighted in some way. (For example, the entry could show up in red type or in boldface type.) If so, your instructor may provide a tutorial on Conditional Formatting or may ask you to refer to Excel Help.

SUMMARY OF KEY RESULTS Section

Your spreadsheet should show the results shown in Figure 6-3. A discussion of line items follows the figure.

	A	B	C	D	E
14	SUMMARY OF KEY RESULTS	2006	2007	2008	2009
15	NET INCOME AFTER TAXES	NA			
16	END-OF-THE-YEAR CASH ON HAND	NA			
17	END-OF-THE-YEAR DEBT OWED	NA			

Figure 6-3 SUMMARY OF KEY RESULTS section

For each year, your spreadsheet should show: (1) net income after taxes; (2) cash on hand at the end of the year; and (3) the debt owed at the end of the year to bankers, as shown in Figure 6-3. These values are all computed elsewhere in the spreadsheet and should be echoed here. These cells should all be formatted for zero decimals.

CALCULATIONS Section

You should calculate various intermediate results that will be used in the income statement and cash flow statement that follows. Calculations, shown in Figure 6-4, are based on inputs and on year 2006 values. When called for, use absolute addressing properly. An explanation of the line items follows the figure.

	A	B	C	D	E
19	CALCULATIONS	2006	2007	2008	2009
20	INTEREST RATE ON DEBT	NA			
21	SELLING PRICE PER UNIT	11.00			
22	NUMBER OF UNITS SOLD	2000000			
23	DIRECT LABOR COST PER UNIT	4.00			
24	OTHER VARIABLE COSTS PER UNIT	3.00			

Figure 6-4 CALCULATIONS section

- INTEREST RATE ON DEBT: If the economy will be *Up*, then the interest rate on debt for the year will be .08 (8%). If the economy will be *Steady*, the interest rate for the year will be .07 (7%). If the economy will be *Down*, the interest rate will be .06 (6%).
- SELLING PRICE PER UNIT: A coffee pot sold for $11 per unit in 2006. If a year's economy will be *Up*, the selling price will be 8% higher than the selling price in the prior year. If a year's economy will be *Steady*, the selling price will be the same as the selling price in the prior year. If a year's economy will be *Down* in a year, the selling price will be 5% less than the selling price in the prior year.
- NUMBER OF UNITS SOLD: Two million coffee pots were sold in 2006. If a year's economy will be *Up*, the number of units sold will increase 3% over the number sold

Case 6

in the prior year. If a year's economy will be *Steady*, the number of units sold will be the same as the number sold in the prior year. If a year's economy will be *Down*, the number of units sold will be 2% less than the number sold in the prior year.

- DIRECT LABOR COST PER UNIT: The direct labor cost per unit sold was $4.00 in 2006. You should enter the change to this direct cost as an input. That input applies to each of the three years. For example, if workers are given a 10% pay raise, then 10% is input as the pay increase factor. The direct labor cost per unit in 2007 would be $4.40. The direct labor cost in 2008 would be $4.40 plus $.44 = $4.84. The direct labor cost in 2009 would be $4.84 + .48 = $5.32.
- OTHER VARIABLE COSTS PER UNIT: This cost was $3.00 per unit in 2006. If work rules are relaxed (*YES* entered for that input), then this cost will go down 2% per year in each of the next three years. For example, in 2007 other variable costs would be 98% of 2006's cost, 2008's would be 98% of 2007's cost, and so on for 2009. On the other hand, if work rules are not relaxed, then other variable costs will increase 10% each year in the next three years; that is, 2007's cost would be $3.30 a unit, 2008's would be $3.30 + .33 = $3.63, and so on for 2009.

INCOME STATEMENT AND CASH FLOW STATEMENT Section

The forecast for net income and cash flow starts with the cash on hand at the beginning of the year. This is followed by the income statement and concludes with the calculation of cash on hand at the year's end. For readability, format cells in this section for zero decimals. Your spreadsheet should look like the ones shown in Figures 6-5 and 6-6. A discussion of line items follows each figure.

	A	B	C	D	E
26	**INCOME STATEMENT AND CASH FLOW STATEMENT**	**2006**	**2007**	**2008**	**2009**
27	BEGINNING-OF-THE-YEAR CASH ON HAND	**NA**			
28					
29	REVENUE	**NA**			
30	RAW MATERIALS COST	**NA**			
31	DIRECT LABOR COST	**NA**			
32	OTHER VARIABLE COSTS	**NA**			
33	FIXED COSTS	**NA**			
34	TOTAL COSTS	**NA**			
35	INCOME BEFORE INTEREST AND TAXES	**NA**			
36	INTEREST EXPENSE	**NA**			
37	INCOME BEFORE TAXES	**NA**			
38	INCOME TAX EXPENSE	**NA**			
39	NET INCOME AFTER TAXES	**NA**			

Figure 6-5 INCOME STATEMENT AND CASH FLOW STATEMENT section

- BEGINNING-OF-THE-YEAR CASH ON HAND: This is the cash on hand at the end of the prior year.
- REVENUE: This is a function of the total number of units sold in a year and the selling price per unit in the year, both of which are calculated values.
- RAW MATERIALS COST: This is a function of the total units sold in a year (a calculation) and the raw material cost per unit (a constant).

- DIRECT LABOR COST: This is a function of the total units sold in a year and the direct labor cost per unit, both of which are calculated values.
- OTHER VARIABLE COSTS: This is a function of the total units sold in a year and the other variable costs per unit, both of which are calculated values.
- FIXED COSTS: This is a constant, which can be echoed here.
- TOTAL COSTS: These are the sum of raw materials, direct labor, other variable costs, and fixed costs.
- INCOME BEFORE INTEREST AND TAXES: This is the difference between revenue and total costs.
- INTEREST EXPENSE: This expense is a function of the debt owed to start the year and the annual interest rate on debt for the year.
- INCOME BEFORE TAXES: This is equal to income before interest and taxes minus interest expense.
- INCOME TAX EXPENSE: This value is zero if the income before taxes is zero or is negative. Otherwise, income tax expense is a function of the year's tax rate and the income before income taxes.
- NET INCOME AFTER TAXES: This is the difference between income before taxes and income tax expense.

Continuing this statement, line items for the year-end cash calculation are discussed. In Figure 6-6, column B is for 2006, column C for 2007, and so on. Year 2006 values are NA except for END-OF-THE-YEAR CASH ON HAND, which is $1,000,000.

	A	B	C	D	E
41	NET CASH POSITION (NCP) BEFORE BORROWING AND REPAYMENT OF DEBT (BEG OF YR CASH + NET INCOME)	**NA**			
42	ADD: BORROWING FROM BANK	**NA**			
43	LESS: REPAYMENT TO BANK	**NA**			
44	EQUALS: END-OF-THE-YEAR CASH ON HAND	1000000			

Figure 6-6 END-OF-THE-YEAR CASH ON HAND section

- NET CASH POSITION (NCP): The NCP at the end of a year equals cash at the beginning of the year, plus the year's net income after taxes.
- ADD: BORROWING FROM BANK: Assume that the company's banker will lend them enough money at year end to get to the minimum cash needed to start the next year. If the NCP is less than this minimum, the company must borrow enough money to get to the minimum. Borrowing increases cash on hand, of course.
- LESS: REPAYMENT TO BANK: If the NCP is more than the minimum cash at the end of a year and there is debt owed, the company must then pay off as much debt as possible (but not take cash below the minimum cash required to start the next year). Repayments reduce cash on hand, of course.
- EQUALS: END-OF-THE-YEAR CASH ON HAND: This equals the NCP plus any borrowings, less any repayments.

DEBT OWED Section

Your spreadsheet body ends with a calculation of debt owed at year end, as shown in Figure 6-7. An explanation of line items follows the figure. Year 2006 values are NA, except that the company owes $10,000,000 at the end of that year.

	A	B	C	D	E
46	DEBT OWED	2006	2007	2008	2009
47	BEGINNING-OF-THE-YEAR DEBT OWED	NA			
48	ADD: BORROWINGS	NA			
49	LESS: REPAYMENT TO BANK	NA			
50	EQUALS: END-OF-THE-YEAR DEBT OWED	10000000			

Figure 6-7 DEBT OWED section

- BEGINNING-OF-THE-YEAR DEBT OWED: Cash owed at the beginning of a year equals cash owed at the end of the prior year.
- ADD: BORROWING: This amount has been calculated elsewhere and can be echoed to this section.
- LESS: REPAYMENT TO BANK: This amount has been calculated elsewhere and can be echoed to this section.
- EQUALS: END-OF-THE-YEAR DEBT OWED: This equals the amount owed at the beginning of a year, plus borrowings in the year, less repayments in the year.

ASSIGNMENT 2 USING THE SPREADSHEET FOR DECISION SUPPORT

You will now complete the case by: (1) using the spreadsheet to gather the data needed to assess the impact of pay rate changes and work rule changes; (2) documenting your recommendation in a memorandum; and (3) making an oral presentation, if your instructor assigns it.

Assignment 2A: Using the Spreadsheet to Gather Data

You have built the spreadsheet to model the labor negotiation. Company management wants to know what the financial results would be in six scenarios. The scenarios are shown in the following shorthand notations:

- ***Steady-One***: *Steady* economy, 8% pay raise, and a work-rules agreement is reached that does relax the work rules. (This is the situation in which the company "buys" a work-rule agreement.)
- ***Steady-Two***: *Steady* economy, 2% pay raise, and work-rules agreement is not reached to relax the rules. (This is the situation in which the company holds down wage-rate increases and then fails to gain union cooperation on work rules.)
- ***Up-One***: *Up* economy, 2% pay raise, and work-rules agreement is not reached to relax the work rules. (This is another situation in which the company does not "buy" a work-rule agreement, but perhaps the *Up* economy overcomes the resulting higher variable costs.)
- ***Up-Two***: *Up* economy, 8% pay raise, and work-rules agreement is reached to relax the work rules. (This is another situation in which the company "buys" a work-rule agreement, but perhaps the *Up* economy overcomes the resulting higher pay-raise costs.)

- ***Down-One***: *Down* economy, 8% pay raise, and work-rules agreement is reached to relax the rules. (This is another situation in which the company "buys" a work-rule cooperation. Do the other variable cost savings overcome the effect of a poor economy?)
- ***Down-Two***: *Down* economy, 2% pay raise, and work-rules agreement is not reached. (This is another situation in which the company does not "buy" a work-rule agreement.)

The likely financial results in each scenario will act as a negotiating guide to management. You will run "what-if" scenarios with the six sets of input values.

- Perform the procedures set forth in Tutorial C to set up and run the Scenario Manager. Record the six possible scenarios. The changing cells are the cells used to input the economic outlook, pay rate increase, and work rule agreement. (*Note*: In the Scenario Manager, you can enter non-contiguous cell ranges as follows: C20…F20, C21, C22. Cell addresses are arbitrary.) The Output cells are the 2009 Summary of Key Results cells.
- When you are done gathering data, print the entire workbook (including the Scenario Summary sheet). Then, save the spreadsheet (File—Save). **POTS.xls** should be the filename.

Assignment 2B: Documenting Your Recommendation in a Memorandum

Open MS Word, and write a brief memorandum to the company's CEO. Your goal is to offer negotiation guidance in each of the three possible economies you modeled. You will want to answer the following questions:

1. If the economy is expected to be *Steady*, what should management strive for in the negotiation—favorable pay rates or work rules? Or, are both options poor?
2. If the economy is expected to be *Up*, what should management do?
3. If the economy is expected to be *Down*, what should management do?

The goal is to have the best 2009 financial results. Here is further guidance on your memorandum:

- Your memorandum should have a proper heading (DATE / TO / FROM / SUBJECT). You might want to use a Word memo template (**File**, click **New**, click **On my computer** in the Templates section, click the **Memos** tab, choose **Contemporary Memo**, and then click **OK**).
- You need not provide background—the CEO is aware of the upcoming negotiations. You should briefly state your analytical method and state the results. Give the CEO your recommendations, keyed to the possible economies.
- Support your recommendation graphically, as your instructor requires: (1) Go back into Excel and put a copy of the Scenario Manager Summary sheet results into the Windows Clipboard. Then, in Word, copy this graphic into the memorandum. (Tutorial C refers to this procedure.) (2) Or, your instructor might want you to make a summary table in Word, based on the Scenario Manager Summary sheet results, after the first paragraph. The procedure for creating a table in Word is described next.

Enter a table into Word, using the following procedure:

1. Select the **Table** menu option, click **Insert**, and then click **Table**.
2. Enter the number of rows and columns.

3. Select **AutoFormat** and choose **TableGrid 1**.
4. Select **OK**, and then select **OK** again.

   Your table should resemble the format of the table shown in Figure 6-8.

Economic Outlook	Pay Rate Increase	Work-Rule Agreement?	2009 Net Income	2009 Cash on Hand	2009 Debt Owed
Steady	.08	Yes			
Steady	.02	No			
Up	.02	No			
Up	.08	Yes			
Down	.08	Yes			
Down	.02	No			

Figure 6-8 Format of table to insert in memorandum

ASSIGNMENT 3 GIVING AN ORAL PRESENTATION

Your instructor might request that you also present your analysis and recommendation in an oral presentation. If so, assume that the CEO has accepted your recommendation. He has asked you to give a presentation explaining your recommendation to the company's senior management and the company's banker. Prepare to explain your analysis and recommendation to the group in 10 minutes or fewer. Use visual aids or handouts that you think are appropriate. Tutorial E has guidance on how to prepare and give an oral presentation.

DELIVERABLES

Assemble the following deliverables for your instructor:

1. Printout of your memorandum
2. Spreadsheet printouts
3. Disk or CD, which should have your Word memo file and your Excel spreadsheet file

Staple the printouts together, with the memorandum on top. If there is more than one *.xls* file on your disk, write your instructor a note, stating the name of your model's *.xls* file.

Freedom National Bank's Buyout Decision

CASE

DECISION SUPPORT USING EXCEL

PREVIEW

Freedom National Bank's management is considering buying a smaller bank, Eastern Oceanic Bank, to gain access to the bank's lucrative credit-card loan portfolio. Freedom National Bank would issue shares of its common stock to buy Eastern Oceanic Bank. Freedom National Bank management wants to know whether their earnings per share will benefit if they buy Eastern Oceanic Bank. In this case, you will use Excel to model the financial implications of the buyout.

PREPARATION

- Review spreadsheet concepts discussed in class and/or in your textbook.
- Complete any exercises that your instructor assigns.
- Complete any part of Tutorial C that your instructor assigns, or refer to it as necessary.
- Review file-saving procedures for Windows programs. These are discussed in Tutorial C.
- Refer to Tutorial E as necessary.

Background

Before looking at Freedom National Bank's decision in depth, let's review some banking basics.

Traditional Bank Operation

Banks make money by charging interest on loans. The borrower pays back the loan over time, and payments include both the principal and interest charge. For example, a person might borrow $15,000 from a bank to buy a car, at 6% interest, to be paid back in a year. The person would pay back an amount equaling the value of the loan ($15,000) and the interest charge (.06 * $15,000 = $900). In a bank's income statement, revenue is the interest earned on such loans. A bank's expenses are for traditional amounts, such as salaries and benefits paid to employees, rent, and so forth.

Where do banks get the money that they lend? Some sources are traditional. For example, some banks sell common stock to the public or bonds in the debt market, just as other kinds of corporations do. However, banks have an additional source of lendable funds: money from depositors. Banks hold depositors' money (in the form of checking and savings accounts), promising to pay out the money on demand to any depositor who wants their money back. Of course, people rarely show up at the teller's window asking to withdraw all of their money, so banks are able to lend out most of the money that they hold on deposit.

Why are people willing to leave their money on deposit with a bank? Because the bank pays interest to the depositors! This interest paid is another expense in the bank's income statement. A bank hopes that the revenue from interest on loans is greater than the total of interest paid to depositors and other expenses.

Bank have assets other than loans. For instance, banks can invest in various kinds of financial instruments, and they will earn interest. A bank's balance sheet will thus show cash, loans, and other assets on the asset side. It will show deposits and other debt as liabilities.

Credit-Card Banks

Freedom National Bank is a very large bank. Its loans are mostly businesses loans and home mortgage loans. Years ago, Freedom National Bank's management decided not to make what are called "credit-card" loans, and now they are regretting that decision.

Credit-card banks issue credit cards to their customers. The cardholder buys goods or services, charging the credit card. The bank pays the seller, and the cardholder pays the bank. If all the charges are not paid at the end of the month, the cardholder owes the unpaid balance to the bank. This balance is a loan from the bank to the cardholder. As with any loan, cardholders pay interest on these loans to the bank.

Interest rates on credit-card loans are generally higher than on other kinds of loans. This is because the loan is based merely on a promise to pay. Unlike a home loan or an auto loan, credit-card debt is "unsecured": If the cardholder fails to pay, the bank has nothing to repossess. In fact, cardholders often fail to pay what they owe, causing losses to the bank. To overcome this expense, banks typically charge much higher interest rates than they do for other kinds of loans.

Not only are credit-card loans riskier than other kinds of loans, but also they require more bookkeeping. The bank must keep track of individual customer purchases and payments—or hire a servicing organization to do so. Because of these two factors, years ago Freedom National Bank opted not to get into the credit-card loan market. A credit-card bank that has loans with people who are likely to repay their debt can be very profitable, however. For such a bank, the interest rates are high and the loan losses are low, so net income is high.

Freedom National Bank's Decision

Eastern Oceanic Bank is a credit-card bank. In fact, all the bank does is issue credit cards and manage their loans. They take in no deposits. The bank is smaller than Freedom National Bank, but it is very profitable.

To gain entry into high quality credit-card loans, Freedom National Bank would very much like to buy Eastern Oceanic Bank—if the price is right. For their part, Eastern Oceanic Bank's management would be very receptive to a takeover offer—if the price is right. The two management teams have been talking. Eastern Oceanic Bank has been driving a hard bargain. Eastern Oceanic has made an offer to sell. You have been called in to make a spreadsheet model of the purchase to see whether the proposed price is reasonable from Freedom National Bank's point of view.

Figures 7-1 and 7-2 present some basic balance sheet data about each company (dollars are in millions). Freedom National Bank data is presented in Figure 7-1; Eastern Oceanic Bank's data is shown in Figure 7-2.

Assets			Liabilities and Owners' Equity		
Account	*Amount*	*Earnings Rate*	*Account*	*Amount*	*Interest Rate*
Cash	$30,000		Deposits	$600,000	.015
Loans	$470,000	.060	Other Debt	$300,000	.020
Other Assets	$500,000	.030	Equity	$100,000	
Total	$1,000,000		Total	$1,000,000	

Figure 7-1 Freedom National Bank Balance Sheet

An explanation of the balance sheet data shown in Figure 7-1 follows:

- *Cash*: This sum is the amount of cash on hand.
- *Loans*: The average interest rate earned on loans has been 6% for Freedom National Bank—that is, interest revenue has been 6% of the value of the loans.
- *Other Assets*: These are investments in securities, the bank's buildings and property, and other assets. The average earnings rate on those assets has been 3%—other revenue earned has been 3% of the value of those assets.

Now let's look at the other side of the balance sheet:

- *Deposits*: The average interest rate paid on *deposits* has been 1.5%.
- *Other Debt*: This is the money borrowed on the open financial markets. Most of this debt is very short term and rolled over quickly. Rates on such debt are low in this era. The average interest rate paid on that debt has been 2%.
- *Equity*: This sum represents the value of the company's common stock outstanding and retained earnings.

Figure 7-2 shows analogous data for Eastern Oceanic Bank. An explanation of the balance sheet follows the figure.

Assets			Liabilities and Owners' Equity	
Account	*Amount*	*Earnings Rate*	*Account*	*Amount*
Cash	$3,000		Other Debt	$30,000
Loans	$50,000	.10	Equity	$50,000
Other Assets	$27,000			
Total	$80,000		Total	$80,000

Figure 7-2 Eastern Oceanic Bank Balance Sheet

Again, *cash* is cash on hand. Eastern Oceanic Bank's credit-card *loans* earn 10%. It's no wonder that Freedom National Bank wants the bank's credit-card loans! In the buyout, Eastern Oceanic Bank's cash and *other assets* would be used to pay off the bank's *other debt*, and the *equity* would be eliminated.

As you can see, Eastern Oceanic Bank is much smaller than Freedom National Bank—$80 billion in assets, versus $1 trillion in assets for Freedom National Bank. To buy Eastern Oceanic Bank, Freedom National Bank would issue 1 billion new shares of its stock, which would be given to Eastern Oceanic Bank shareholders in return for their stock, which would then be retired.

A billion shares of stock at first might seem like a lot of stock shares. Freedom National Bank's management would hope for two things to happen in the combined bank: (1) They could "cross-sell" new kinds of loans to credit-card borrowers; and (2) the credit-card loan portfolio would grow.

In "cross-selling," the bank would make new auto loans, home loans, and other kinds of loans to holders of credit cards. The assumption is that these new loans would not have been made if the card accounts had not been bought. On the surface, this does seem like a plausible strategy—after all, why wouldn't a cardholder think of Freedom National Bank first for all their borrowing needs? Freedom National Bank's management thinks they can make cross-selling work, but as an analyst, you should know that other banks have failed at this sort of strategy when trying to make a buyout work. So, one question for the combined bank is this: *Can cross-selling work for Freedom National Bank?*

Eastern Oceanic Bank's credit-card loan portfolio has grown about 10% in most years. In recent years however, loan growth has slowed at all the credit-card banks. The primary reason seems to be the "real estate bubble": Mortgage rates are low and house values keep going up. People refinance their home mortgage, taking out some cash as they do so—in effect, they are borrowing against the value of their house. They then use the cash to pay off high-interest-rate debts, including credit-card loans. In fact, in the past year, the value of Eastern Oceanic Bank's loans actually declined! Freedom National Bank's management expects loan growth to resume, but clearly one question for the combined bank is this: *What will be the growth rate for credit-card loans?*

The ultimate question for Freedom National Bank management is this: *Is this purchase price right?* Freedom National Bank management wants their company's earnings per share (EPS) to go up as the result of the buyout. EPS is the ratio of a company's net income after taxes ("earnings") to the number of common stock shares outstanding. The EPS of the combined bank should be greater than what Freedom National Bank's EPS would be without the buyout.

A model of the projected Freedom National Bank income statement with and without Eastern Oceanic Bank is needed. You have been called in to create the model. You are asked to make a projected 2007–2009 forecast in Excel. Certain scenarios are under consideration, and your spreadsheet would handle them in the Scenario Manager.

ASSIGNMENT 1 CREATING A SPREADSHEET FOR DECISION SUPPORT

In this assignment, you will produce a spreadsheet that models the business decision. Then, in Assignment 2, you will write a memorandum to the bank's management that explains your recommended action. In addition, in Assignment 3, you will be asked to prepare an oral presentation of your analysis and recommendation.

First, you will create the spreadsheet model of the buyout. The model is an analysis for the years 2007–2009. You will be given some hints on how each section should be set up before entering cell formulas. Your spreadsheet should have the sections that follow:

- CONSTANTS
- INPUTS
- SUMMARY OF KEY RESULTS
- CALCULATIONS
- INCOME STATEMENT AND CASH FLOW STATEMENT

A discussion of each section follows. *The spreadsheet skeleton is available to you, so you need not type in the skeleton if you do not want to do so.* To access the spreadsheet, go to your Data files, select Case 7, and then select **FREEDOM.xls**.

CONSTANTS Section

Your spreadsheet should have the constants shown in Figure 7-3. An explanation of the line items follows the figure.

	A	B	C	D	E
1	FREEDOM NATIONAL BANK PURCHASE ANALYSIS				
2					
3	CONSTANTS	2006	2007	2008	2009
4	TAX RATE	NA	0.33	0.33	0.33
5	CASH NEEDED TO START THE YEAR	NA	30000000000	30000000000	30000000000
6	EARNINGS RATE -- OTHER ASSETS	NA	0.0300	0.0310	0.0320
7	EXPECTED FREEDOM LOAN GROWTH	NA	0.0200	0.0200	0.0200
8	EARNINGS RATE -- FREEDOM LOANS	NA	0.0600	0.0620	0.0640
9	AVERAGE PAY PER EMPLOYEE	NA	85000	86000	87000
10	OTHER EXPENSES -- FREEDOM	NA	8000000000	8200000000	8300000000
11	EARNINGS RATE -- CREDIT CARD LOANS	NA	0.1000	0.1000	0.1000
12	OTHER EARNING ASSETS	NA	500000000000	510000000000	520000000000
13	DEPOSITS OWED	NA	600000000000	610000000000	620000000000
14	INTEREST RATE ON DEPOSITS OWED	NA	0.0150	0.0155	0.0160
15	INTEREST RATE ON DEBT OWED	NA	0.0200	0.0205	0.0210

Figure 7-3 CONSTANTS section

- TAX RATE: The expected tax rate on income before taxes for 2007–2009 is shown as a constant 33%.
- CASH NEEDED TO START THE YEAR: The bankers say they need to start each year with a certain amount of cash. If cash falls below that amount, the bank will have to borrow in the financial markets.

Case 7

- EARNINGS RATE—OTHER ASSETS: Revenue earned on "other assets" is expected to be 3% of the value of those assets in 2007, increasing slightly each year thereafter.
- EXPECTED FREEDOM LOAN GROWTH: Freedom National Bank's current loan portfolio is expected to grow 2% a year in each year. (This rate applies to non-credit-card loans.)
- EARNINGS RATE—FREEDOM LOANS: Revenue earned on Freedom National Bank loans is expected to be 6% of the value of those loans in 2007, increasing slightly each year thereafter. (This rate applies to non-credit-card loans.)
- AVERAGE PAY PER EMPLOYEE: An employee is expected to cost the bank $85,000 in salary and benefits in 2007, increasing in future years. This rate applies to employees of both banks.
- OTHER EXPENSES—FREEDOM: The bank incurs costs for computer services, networks, rent, and so forth. The value of other expenses is expected to be $8 billion in 2007, increasing in the next two years. This amount is for Freedom National Bank operations only.
- EARNINGS RATE—CREDIT CARD LOANS: Revenue earned on Eastern Oceanic Bank's credit-card loans is expected to be 10% of the value of those loans in 2007–2009.
- OTHER EARNING ASSETS: Other Freedom National Bank assets are expected to be $500 billion in 2007, increasing each year.
- DEPOSITS OWED: Freedom National Bank is expected to owe $600 billion to depositors in 2007, increasing each year.
- INTEREST RATE ON DEPOSITS OWED: Freedom National Bank pays depositors 1.5% interest. This rate is expected to increase in succeeding years.
- INTEREST RATE ON DEBT OWED: Freedom National Bank expects to pay 2% on debt owed in 2007, with the rate increasing in each of the next two years.

INPUTS Section

Your spreadsheet should have the inputs shown in Figure 7-4. An explanation of the line items follows the figure.

	A	B	C	D	E
17	INPUTS	2006	2007	2008	2009
18	BUY BANK? (B = BUY, DB = DON'T BUY)	NA		NA	NA
19	CROSS-SELLING WORKS? (Y = YES, N = NO)	NA		NA	NA
20	CREDIT CARD LOAN GROWTH	NA			

Figure 7-4 INPUTS section

- BUY BANK? If Eastern Oceanic Bank will be bought, enter a *B* for *Buy*; otherwise, enter *DB* for *Don't Buy*. The value applies to all three analysis years.
- CROSS-SELLING WORKS? If cross-selling is expected to be successful, enter a *Y* for *Yes*; otherwise, enter an *N* for *No*. The value applies to all three analysis years.
- CREDIT CARD LOAN GROWTH: For each year, enter the rate of growth in the credit-card loan portfolio. For example, if loans are expected to increase 5% in value, then enter .05. If loans are expected to decrease 2% in value, enter –.02. A value for each year is entered. For example, the sequence .01, .02, .03 would indicate a slowly growing loan portfolio in the three years.

Your instructor might tell you to apply Conditional Formatting to the input cells, so out-of-bounds values are highlighted in some way. (For example, the entry shows up in red type or in boldface type.) If so, your instructor may provide a tutorial on Conditional Formatting or might ask you to refer to Excel Help.

SUMMARY OF KEY RESULTS Section

For each year, your spreadsheet should show earnings per share as shown in Figure 7-5.

	A	B	C	D	E
22	SUMMARY OF KEY RESULTS	2006	2007	2008	2009
23	EARNINGS PER SHARE	NA			

Figure 7-5 SUMMARY OF KEY RESULTS section

Earnings per share is net income after taxes divided by the number of shares outstanding. The EPS values are calculated elsewhere in the spreadsheet and then echoed to this section. These cells should all be formatted for two decimals.

CALCULATIONS Section

You should calculate various intermediate results that will be used in the sections that follow. Calculations, shown in Figure 7-6, are based on inputs and/or on year 2006 values. When called for, use absolute addressing properly. Cells in this section should be formatted for zero decimals. An explanation of the line items follows the figure.

	A	B	C	D	E
25	CALCULATIONS	2006	2007	2008	2009
26	CREDIT CARD LOANS	50000000000			
27	FREEDOM LOANS	470000000000			
28	OTHER REVENUE	NA			
29	TOTAL NUMBER OF EMPLOYEES	NA			
30	SHARES OUTSTANDING	NA			
31	OTHER EXPENSES	NA			

Figure 7-6 CALCULATIONS section

- CREDIT CARD LOANS: The amount of the loan portfolio will be a function of the value of the portfolio to start a year and the credit-card loan growth rate, which is an input value. For example, if 10% growth is expected, there will be $55 billion in loans in 2007, and there will be 10% more than that in 2008, and so on for 2009.
- FREEDOM LOANS: Several factors must be considered:
  1. If Eastern Oceanic Bank is *not* bought, the amount of Freedom National Bank's loans will be a function of the value of the loans to start the year and the expected Freedom National Bank growth rate, which is a constant.
  2. If Eastern Oceanic Bank is bought but cross-selling is not expected to work, then (again) the amount of Freedom National Bank loans will be a function of the value of the loans to start the year and the expected Freedom National Bank growth rate, which is a constant.
  3. If Eastern Oceanic Bank is bought and cross-selling is expected to work, then the amount of Freedom National Bank loans will be a function of the value of the loans to start the year and *twice* the expected Freedom National Bank growth rate, which is a constant.

- OTHER REVENUE: This value is a function of other earnings assets and the earnings rate on other assets. Both these values are constants.
- TOTAL NUMBER OF EMPLOYEES: If the bank is bought, total employees will be 200,000; otherwise, total employees will be 180,000 people in each year.
- SHARES OUTSTANDING: If the bank is not bought, 3.75 billion shares of Freedom National Bank common stock will be outstanding. If the bank is bought, 4.75 billion shares of Freedom National Bank common stock will be outstanding.
- OTHER EXPENSES: If the bank is not bought, this amount is the value shown in constants. If the bank is bought, other expenses are projected to be 120% of the constant value for the year.

INCOME STATEMENT AND CASH FLOW STATEMENT Section

The forecast for net income and cash flow starts with the cash on hand at the beginning of the year. This is followed by the income statement and concludes with the calculation of cash on hand at year-end. For readability, format cells in this section for zero decimals. Your spreadsheet should look like that shown in Figures 7-7 and 7-8. A discussion of line items follows each figure.

	A	B	C	D	E
33	**INCOME STATEMENT AND CASH FLOW STATEMENT**	**2006**	**2007**	**2008**	**2009**
34	BEGINNING-OF-THE-YEAR CASH ON HAND	**NA**			
35					
36	REVENUE:	--	--	--	--
37	INTEREST ON CREDIT CARD LOANS	**NA**			
38	INTEREST ON OTHER LOANS	**NA**			
39	OTHER REVENUE	**NA**			
40	TOTAL REVENUE	**NA**			
41	EXPENSES:	--	--	--	--
42	EMPLOYMENT EXPENSE	**NA**			
43	OTHER EXPENSE	**NA**			
44	TOTAL EXPENSES	**NA**			
45	INCOME BEFORE INTEREST AND TAXES	**NA**			
46	INTEREST EXPENSE	--	--	--	--
47	INTEREST PAID TO DEPOSITORS	**NA**			
48	INTEREST PAID TO DEBT HOLDERS	**NA**			
49	TOTAL INTEREST EXPENSE	**NA**			
50	INCOME BEFORE TAXES	**NA**			
51	INCOME TAX EXPENSE	**NA**			
52	NET INCOME AFTER TAXES	**NA**			

Figure 7-7 INCOME STATEMENT AND CASH FLOW STATEMENT section

- BEGINNING-OF-THE-YEAR CASH ON HAND: This is the cash on hand at the end of the prior year.
- INTEREST ON CREDIT CARD LOANS: This revenue amount is a function of the earnings rate (a constant) and the value of the credit-card loans (a calculation).
- INTEREST ON OTHER LOANS: This revenue amount is a function of the earnings rate (a constant) and the value of the other Freedom National Bank loans (a calculation).
- OTHER REVENUE: This amount is a calculation that can be echoed here.
- TOTAL REVENUE: This is the sum of interest on credit-card loans, interest on other loans, and other revenue.
- EMPLOYMENT EXPENSE: This is a function of the number of employees (a calculation) and the average pay rate per employee (a constant).

- OTHER EXPENSE: This amount is a calculation that can be echoed here.
- TOTAL EXPENSES: These are the sum of employment expense and other expenses.
- INCOME BEFORE INTEREST AND TAXES: This is the difference between total revenue and total expenses.
- INTEREST PAID TO DEPOSITORS: This is a function of total deposits owed and the interest rate on deposits. Both amounts are constants.
- INTEREST PAID TO DEBT HOLDERS: This is a function of the interest rate paid to debt holders and the amount of debt owed at the beginning of the year.
- TOTAL INTEREST EXPENSE: This is the sum of interest paid to depositors and interest paid to other debt holders.
- INCOME BEFORE TAXES: This is the difference between income before interest and taxes and total interest expense.
- INCOME TAX EXPENSE: This value is zero if the income before taxes is zero or is negative. Otherwise, income tax expense is a function of the year's tax rate and the income before income taxes.
- NET INCOME AFTER TAXES: This is the difference between income before taxes and income tax expense.

Continuing this statement, line items for the year-end cash calculation are discussed. In Figure 7-8, column B is for 2006, column C for 2007, and so on. Year 2006 values are mostly NA, except that END-OF-THE-YEAR CASH ON HAND was $30 billion.

	A	B	C	D	E
54	NET CASH POSITION (NCP) BEFORE BORROWING AND REPAYMENT OF DEBT (BEG OF YR CASH + NET INCOME)	NA			
55	ADD: BORROWING	NA			
56	LESS: REPAYMENT OF DEBT	NA			
57	EQUALS: END-OF-THE-YEAR CASH ON HAND	30000000000			

Figure 7-8 END-OF-THE-YEAR CASH ON HAND section

- NET CASH POSITION (NCP): The NCP at the end of a year equals cash at the beginning of the year, plus the year's net income after taxes.
- ADD: BORROWING: Assume that the financial markets will provide enough money at year-end to get to the minimum cash needed to start the next year. If the NCP is less than this minimum, Freedom National Bank must borrow enough money in the financial markets to get to the minimum. Borrowing increases cash on hand, of course.
- LESS: REPAYMENT OF DEBT: If the NCP is more than the minimum cash at the end of a year and there is debt owed, the bank will then pay off as much debt as possible (but not take cash below the minimum cash required to start the next year). Repayments reduce cash on hand, of course.
- EQUALS: END-OF-THE-YEAR CASH ON HAND: This equals the NCP plus any borrowings, less any repayments.

Case 7

DEBT OWED Section

Your spreadsheet body continues with a calculation of debt owed at year-end, as shown in Figure 7-9. An explanation of line items follows the figure. Year 2006 values are mostly NA, except that the Freedom National Bank owes $300 billion at the end of that year.

	A	B	C	D	E
59	DEBT OWED	2006	2007	2008	2009
60	BEGINNING-OF-THE-YEAR DEBT OWED	NA			
61	ADD: BORROWING	NA			
62	LESS: REPAYMENT OF DEBT	NA			
63	EQUALS: END-OF-THE-YEAR DEBT OWED	300000000000			

Figure 7-9 DEBT OWED section

- BEGINNING-OF-THE-YEAR DEBT OWED: Cash owed at the beginning of a year equals cash owed at the end of the prior year.
- ADD: BORROWING: This amount has been calculated elsewhere and can be echoed to this section.
- LESS: REPAYMENT OF DEBT: This amount has been calculated elsewhere and can be echoed to this section.
- EQUALS: END-OF-THE-YEAR DEBT OWED: This equals the amount owed at the beginning of a year, plus borrowings in the year, less repayments in the year.

EARNINGS PER SHARE Section

Your spreadsheet body ends with a calculation of earnings per share for the year, shown in Figure 7-10. An explanation of line items follows the figure. Year 2006 values are NA. The earnings per share cells should be formatted for three decimal places.

	A	B	C	D	E
65	EARNINGS PER SHARE	2006	2007	2008	2009
66	NET INCOME AFTER TAXES	NA			
67	SHARES OUTSTANDING	NA			
68	EARNINGS PER SHARE	NA			

Figure 7-10 EARNINGS PER SHARE section

- NET INCOME AFTER TAXES: This amount has been calculated elsewhere and can be echoed here.
- SHARES OUTSTANDING: This amount has been calculated elsewhere and can be echoed here.
- EARNINGS PER SHARE: This equals net income after taxes for the year divided by shares outstanding for the year. For example, if net income in a year is $16 billion and there are 4.75 billion shares outstanding: EPS = $16 / 4.75 = $3.368 for the year.

Assignment 2 Using the Spreadsheet for Decision Support

You will now complete the case by: (1) using the spreadsheet to gather the data needed to assess the impact of the acquisition; (2) documenting your recommendation in a memorandum; and (3) making an oral presentation, if your instructor assigns it.

Assignment 2A: Using the Spreadsheet to Gather Data

You have built the spreadsheet to model the acquisition. Freedom National Bank management wants to know what the 2009 earnings per share (EPS) would be in seven scenarios. The scenarios are shown in the following shorthand notations:

- ***No Buy***: The bank is not bought. Enter *DB* (Don't Buy). Enter an *N* (No) value for cross-selling. Enter a zero for credit-card loan growth. This scenario shows Freedom National Bank's EPS without Eastern Oceanic Bank.
- ***Buy-1***: The bank is bought, cross-selling works, but there is zero credit-card loan growth.
- ***Buy-2***: The bank is bought, cross-selling works, and there is credit-card loan growth: 2% in 2007, 3% in 2008, and 4% in 2009. Clearly, this is an optimistic scenario for the bank.
- ***Buy-3***: The bank is bought, cross-selling works, but credit-card loan growth is negative: –2% in each of 2007, 2008, and 2009.
- ***Buy-4***: The bank is bought, cross-selling does not work, and there is zero credit-card loan growth. This is clearly not an optimistic scenario.
- ***Buy-5***: The bank is bought, cross-selling does not work, but there is credit-card loan growth: 2% in 2007, 3% in 2008, and 4% in 2009.
- ***Buy-6***: The bank is bought, cross-selling does not work, and credit-card loan growth is negative: –2% in each of 2007, 2008, and 2009. This is another pessimistic scenario.

Freedom National Bank management wants to know the answer to these three questions:

1. How important is cross-selling? If cross-selling works, does it matter if there is credit-card growth?
2. How important is credit-card growth? If growth is zero or negative, can EPS still improve?
3. Under what circumstances would it be best for Freedom National Bank not to buy Eastern Oceanic Bank?

The likely EPS results in each scenario will act as a decision-making guide to management. You will run "what-if" scenarios with the seven sets of input values.

- Perform the procedures set forth in Tutorial C to set up and run the Scenario Manager. Record the seven possible scenarios. The changing cells are the cells used to input the buying decision, cross-selling outlook, and loan growth percentages. (*Note*: In the Scenario Manager, you can enter noncontiguous cell ranges as follows: C20…F20, C21, C22—cell addresses are arbitrary.) The Output cell is the 2009 Summary of Key Results EPS cell.
- When you are done gathering data, print the entire workbook (including the Scenario Summary sheet). Then, save the spreadsheet (File—Save). **FREEDOM.xls** should be the filename.

Assignment 2B: Documenting Your Recommendation in a Memorandum

Open MS Word, and write a brief memorandum to the Freedom National Bank CEO. Your goal is to offer financial guidance. You should try to answer the three questions stated previously. Use the Scenario Manager results to support your conclusions. In addition, render your overall conclusion: Should Freedom National Bank buy the credit-card bank? Use the Scenario Manager results to support that conclusion as well.

- Your memorandum should have a proper heading (DATE / TO / FROM / SUBJECT). You might want to use a Word memo template (**File**, click **New**, click **On my computer** in the Templates section, click the **Memos** tab, choose **Contemporary Memo**, and then click **OK**).
- You need not provide background—the CEO is aware of the upcoming negotiations. You should briefly state your analytical method and state the results. Give the CEO your recommendations, keyed to the possible economies.
- Support your recommendation graphically, as your instructor requires: (1) Go back into Excel, and put a copy of the Scenario Manager Summary sheet results into the Windows Clipboard. Then, in Word, copy this graphic into the memorandum (Tutorial C refers to this procedure). (2) Or, your instructor might want you to make a summary table in Word, based on the Scenario Manager Summary sheet results, after the first paragraph. The procedure for creating a table in Word is described next.

Enter a table into Word, using the following procedure:

1. Select the **Table** menu option, click **Insert**, and then click **Table**.
2. Enter the number of rows and columns.
3. Select **AutoFormat** and choose **TableGrid 1**.
4. Select **OK**, and then select **OK** again.

Your table should resemble the format of the table shown in Figure 7-11.

	Buy?	**Cross-sell Works?**	**2007 Loan Growth**	**2008 Loan Growth**	**2009 Loan Growth**	**2009 EPS**
No Buy	No	—	—	—	—	
Buy 1	Yes	Yes	0	0	0	
Buy 2	Yes	Yes	.02	.03	.04	
Buy 3	Yes	Yes	–.02	–.02	–.02	
Buy 4	Yes	No	0	0	0	
Buy 5	Yes	No	.02	.03	.04	
Buy 6	Yes	No	–.02	–.02	–.02	

Figure 7-11 Format of table to insert in memorandum

ASSIGNMENT 3 GIVING AN ORAL PRESENTATION

Your instructor might request that you also present your analysis and recommendations in an oral presentation. If so, assume that the Freedom National Bank CEO has accepted your recommendation. She has asked you to give a presentation explaining your recommendation to the bank's senior management and the bank's financial market lenders. Prepare to explain your analysis and recommendation to the group in 10 minutes or fewer. Use visual aids or handouts that you think are appropriate. Tutorial E has guidance on how to prepare and give an oral presentation.

DELIVERABLES

Assemble the following deliverables for your instructor:

1. Printout of your memorandum
2. Spreadsheet printouts
3. Disk or CD, which should have your Word memo file and your Excel spreadsheet file

Staple the printouts together, with the memorandum on top. If there is more than one *.xls* file on your disk, write your instructor a note, stating the name of your *.xls* file for this case.

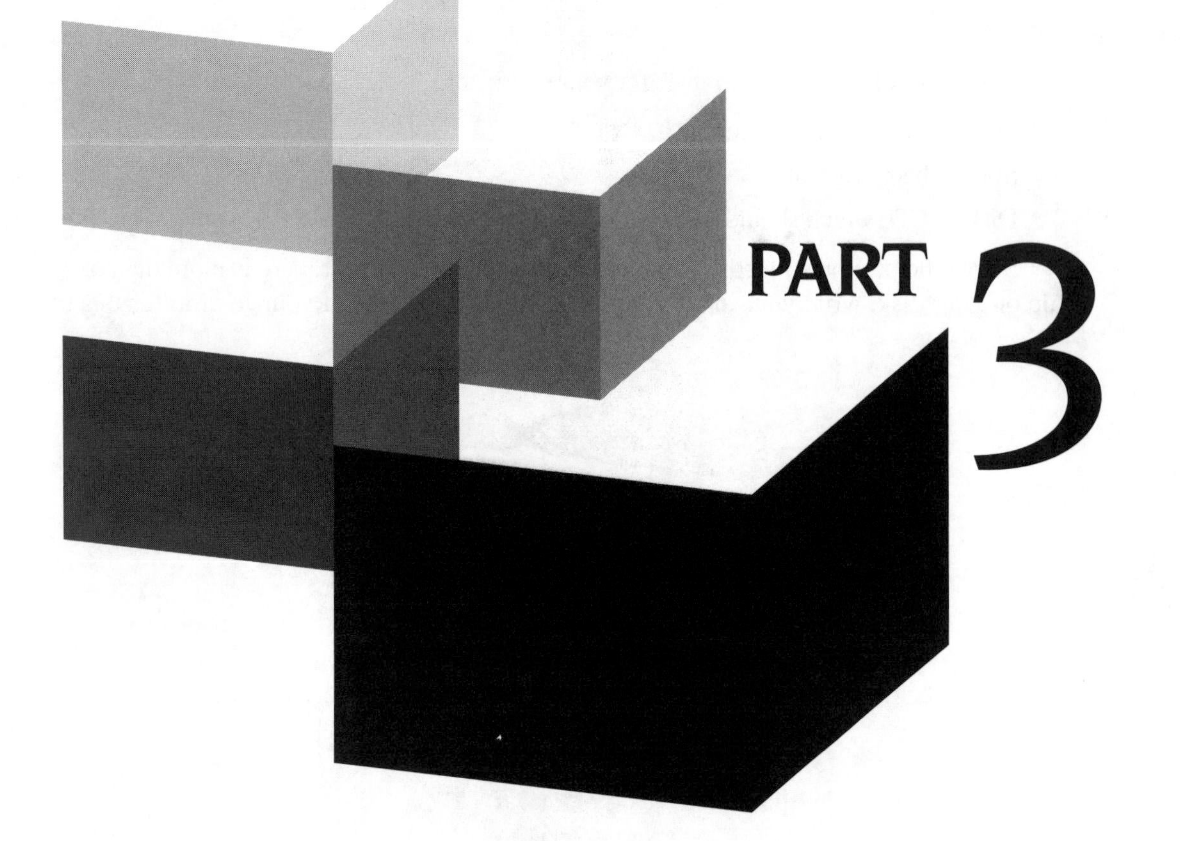

PART 3

Decision Support Cases Using the Excel Solver

Building a Decision Support System Using the Excel Solver

Decision Support Systems (DSS) help people to make decisions. (The nature of DSS programs is discussed in Tutorial C.) Tutorial D teaches you how to use the Solver, one of the Excel built-in decision support tools.

For some business problems, decision makers want to know the best, or optimal, solution. Usually this means maximizing a variable (for example, net income) or minimizing another variable (for example, total costs). This optimization is subject to constraints, which are rules that must be observed when solving a problem. The Solver computes answers to such optimization problems.

This tutorial has four sections:

1. **Using the Excel Solver** In this section, you'll learn how to use the Solver in decision making. As an example, you use the Solver to create a production schedule for a sporting goods company. This schedule is called the Base Case.
2. **Extending the Example** In this section, you'll test what you've learned about using the Solver as you modify the sporting goods company's production schedule. This is called the Extension Case.
3. **Using the Solver on a New Problem** In this section, you'll use the Solver on a new problem.
4. **Trouble-Shooting the Solver** In this section, you'll learn how to overcome problems you might encounter when using the Solver.

Tutorial C has some guidance on basic Excel concepts, such as formatting cells and using functions, such as =IF(). Refer to Tutorial C for a review of such topics.

Using the Excel Solver

Suppose that a company must set a production schedule for its various products, each of which has a different profit margin (selling price less costs). At first, you might assume that the company will maximize production of all profitable products to maximize net income. However, a company typically cannot make and sell an unlimited number of its products because of constraints.

One constraint affecting production is the "shared resource problem." For example, several products in a manufacturer's line might require the same raw materials, which are in limited supply. Similarly, the manufacturer might require the same machines to make several of its products. In addition, there might also be a limited pool of skilled workers available to make the products.

In addition to production constraints, sometimes management's policies impose constraints. For example, management might decide that the company must have a broader product line. As a consequence, a certain production quota for several products must be met, regardless of profit margins.

Thus, management must find a production schedule that will maximize profit, given the constraints. Optimization programs like the Solver look at each combination of products, one after the other, ranking each combination by profitability. Then the program reports the most profitable combination.

To use the Solver, you'll set up a model of the problem, including the factors that can vary, the constraints on how much they can vary, and the goal you are trying to maximize (usually net income) or minimize (usually total costs). The Solver then computes the best solution.

Setting Up a Spreadsheet Skeleton

Suppose that your company makes two sporting goods products—basketballs and footballs. Assume that you will sell all the balls you produce. To maximize net income, you want to know how many of each kind of ball to make in the coming year.

Making each kind of ball requires a certain (and different) number of hours, and each ball has a different raw materials cost. Because you have only a limited number of workers and machines, you can devote a maximum of 40,000 hours to production. This is a shared resource. You do not want that resource to be idle, however. Downtime should be no more than 1,000 hours in the year, so machines should be used for at least 39,000 hours.

Marketing executives say you cannot make more than 60,000 basketballs and may not make fewer than 30,000. Furthermore, they say that you must make at least 20,000 footballs but not more than 40,000. Marketing says the ratio of basketballs to footballs produced should be between 1.5 and 1.7—that is, more basketballs than footballs, but within limits.

What would be the best production plan? This problem has been set up in the Solver. The spreadsheet sections are discussed in the pages that follow.

AT THE KEYBOARD

Start out by saving the blank spreadsheet as **SPORTS1.xls**. Then you should enter the skeleton and formulas as they are discussed.

CHANGING CELLS Section

The **CHANGING CELLS** section contains the variables the Solver is allowed to change while it looks for the solution to the problem. Figure D-1 shows the skeleton of this spreadsheet section and the values that you should enter. An analysis of the line items follows the figure.

	A	B
1	SPORTING GOODS EXAMPLE	
2	CHANGING CELLS	
3	NUMBER OF BASKETBALLS	1
4	NUMBER OF FOOTBALLS	1

Figure D-1 CHANGING CELLS section

- The changing cells are for the number of basketballs and footballs to be made and sold. The changing cells are like input cells, except Solver (not you) plays "what-if" with the values, trying to maximize or minimize some value (in this case, maximize net income).
- Note that some number should be put in the changing cells each time before the Solver is run. It's customary to put the number 1 into the changing cells (as shown). Solver will change these values when the program is run.

CONSTANTS Section

Your spreadsheet should also have a section for values that will not change. Figure D-2 shows a skeleton of the **CONSTANTS** section and the values you should enter. A discussion of the line items follows the figure.

You should use Format—Cells—Number to set the constants range to two decimal places.

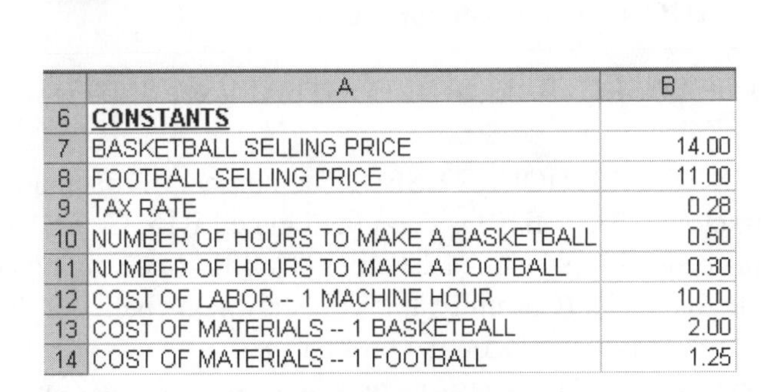

	A	B
6	CONSTANTS	
7	BASKETBALL SELLING PRICE	14.00
8	FOOTBALL SELLING PRICE	11.00
9	TAX RATE	0.28
10	NUMBER OF HOURS TO MAKE A BASKETBALL	0.50
11	NUMBER OF HOURS TO MAKE A FOOTBALL	0.30
12	COST OF LABOR -- 1 MACHINE HOUR	10.00
13	COST OF MATERIALS -- 1 BASKETBALL	2.00
14	COST OF MATERIALS -- 1 FOOTBALL	1.25

Figure D-2 CONSTANTS section

- The SELLING PRICE for one basketball and for one football is shown.
- The TAX RATE is the rate applied to income before taxes to compute income tax expense.
- The NUMBER OF MACHINE HOURS needed to make a basketball and a football is shown. Note that a ball-making machine can produce two basketballs in an hour.
- COST OF LABOR: A ball is made by a worker using a ball-making machine. A worker is paid $10 for each hour he or she works at a machine.
- COST OF MATERIALS: The costs of raw materials for a basketball and football are shown.

Notice that the profit margins (selling price less costs of labor and materials) for the two products are not the same. They have different selling prices and different inputs (raw materials, hours to make) and the inputs have different costs per unit. Also note that you cannot tell from the data how many hours of the shared resource (machine hours) will be devoted to basketballs and how many to footballs, because you don't know in advance how many basketballs and footballs will be made.

CALCULATIONS Section

In the **CALCULATIONS** section, you will calculate intermediate results that (1) will be used in the spreadsheet body, and/or (2) will be used as constraints. First, use Format—Cells—Number to set the calculations range to two decimal places. Figure D-3 shows the skeleton and formulas that you should enter. A discussion of the cell formulas follows the figure.

Cell widths are changed here merely to show the formulas—you need not change the width.

	A	B
16	CALCULATIONS	
17	RATIO OF BASKETBALLS TO FOOTBALLS	=B3/B4
18	TOTAL BASKETBALL HOURS USED	=B3*B10
19	TOTAL FOOTBALL HOURS USED	=B4*B11
20	TOTAL MACHINE HOURS USED (BB + FB)	=B18+B19

Figure D-3 CALCULATIONS section cell formulas

- The RATIO OF BASKETBALLS TO FOOTBALLS (cell B17) will be needed in a constraint.
- TOTAL BASKETBALL HOURS USED: The number of machine hours needed to make all basketballs (B3 * B10) is computed in cell B18. Cell B10 has the constant for the hours needed to make one basketball. Cell B3 (a changing cell) has the number of basketballs made. (Currently, this cell shows one ball, but that number will change when the Solver works on the problem.)
- TOTAL FOOTBALL HOURS USED: The number of machine hours needed to make all footballs is calculated similarly, in cell B19.
- TOTAL MACHINE HOURS USED (BB + FB): The number of hours needed to make both kinds of balls (cell B20) will be a constraint; this value is the sum of the hours just calculated for footballs and basketballs.

Notice that constants in the Excel cell formulas in Figure D–3 are referred to by their cell addresses. Use the cell address of a constant rather than hard-coding a number in the Excel expression: If the number must be changed later, you only have to change it in the CONSTANTS section cell, not in every cell formula in which you used the value.

Notice that you do not calculate the amounts in the changing cells (here, the number of basketballs and footballs to produce). The Solver will compute those numbers. Also notice that you can use the changing cell addresses in your formulas. When you do that, you assume the Solver has put the optimal values in each changing cell; your expression makes use of that number.

Figure D-4 shows the values after Excel evaluates the cell formulas (with 1s in the changing cells):

	A	B
16	**CALCULATIONS**	
17	RATIO OF BASKETBALLS TO FOOTBALLS	1.00
18	TOTAL BASKETBALL HOURS USED	0.50
19	TOTAL FOOTBALL HOURS USED	0.30
20	TOTAL MACHINE HOURS USED (BB + FB)	0.80

Figure D-4 CALCULATIONS section cell values

INCOME STATEMENT Section

The target value is calculated in the spreadsheet body in the INCOME STATEMENT section. This is the value that the Solver is expected to maximize or minimize. The spreadsheet body can take any form. In this textbook's Solver cases, the spreadsheet body will be an income statement. Figure D-5 shows the skeleton and formulas that you should enter. A discussion of the line-item cell formulas follows the figure.

NOTE

Income statement cells were formatted for two decimal places.

	A	B
22	**INCOME STATEMENT**	
23	BASKETBALL REVENUE (SALES)	=B3*B7
24	FOOTBALL REVENUE (SALES)	=B4*B8
25	TOTAL REVENUE	=B23+B24
26	BASKETBALL MATERIALS COST	=B3*B13
27	FOOTBALL MATERIALS COST	=B4*B14
28	COST OF MACHINE LABOR	=B20*B12
29	TOTAL COST OF GOODS SOLD	=SUM(B26:B28)
30	INCOME BEFORE TAXES	=B25-B29
31	INCOME TAX EXPENSE	=IF(B30<=0,0,B30*B9)
32	NET INCOME AFTER TAXES	=B30-B31

Figure D-5 INCOME STATEMENT section cell formulas

- REVENUE (cells B23 and B24) equals the number of balls times the respective unit selling price. The number of balls is in the changing cells, and the selling prices are constants.
- MATERIALS COST (cells B26 and B27) follows a similar logic: number of units times unit cost.
- COST OF MACHINE LABOR is the calculated number of machine hours times the hourly labor rate for machine workers.
- TOTAL COST OF GOODS SOLD is the sum of the cost of materials and the cost of labor.

- This is the logic of income tax expense: If INCOME BEFORE TAXES is less than or equal to zero, the tax is zero; otherwise, the income tax expense equals the tax rate times income before taxes. An =IF() statement is needed in cell B31.

Excel evaluates the formulas. Figure D-6 shows the results (assuming 1s in the changing cells):

	A	B
22	**INCOME STATEMENT**	
23	BASKETBALL REVENUE (SALES)	14.00
24	FOOTBALL REVENUE (SALES)	11.00
25	TOTAL REVENUE	25.00
26	BASKETBALL MATERIALS COST	2.00
27	FOOTBALL MATERIALS COST	1.25
28	COST OF MACHINE LABOR	8.00
29	TOTAL COST OF GOODS SOLD	11.25
30	INCOME BEFORE TAXES	13.75
31	INCOME TAX EXPENSE	3.85
32	NET INCOME AFTER TAXES	9.90

Figure D-6 INCOME STATEMENT section cell values

Constraints

Constraints are rules that the Solver must observe when computing the optimal answer to a problem. Constraints will need to refer to calculated values, or to values in the spreadsheet body. Therefore, you must build those calculations into the spreadsheet design, so they are available to your constraint expressions. (There is no section on the face of the spreadsheet for constraints. You'll use a separate window to enter constraints.)

Figure D-7 shows the English and Excel expressions for the basketball and football production problem constraints. A discussion of the constraints follows the figure.

Expression in English	Excel Expression
TOTAL MACHINE HOURS >= 39000	B20 >= 39000
TOTAL MACHINE HOURS <= 40000	B20 <= 40000
MIN BASKETBALLS = 30000	B3 >= 30000
MAX BASKETBALLS = 60000	B3 <= 60000
MIN FOOTBALLS = 20000	B4 >= 20000
MAX FOOTBALLS = 40000	B4 <= 40000
RATIO BBs TO FBs-MIN = 1.5	B17 >= 1.5
RATIO BBs TO FBs-MAX = 1.7	B17 <= 1.7
NET INCOME MUST BE POSITIVE	B32 >= 0

Figure D-7 Solver Constraint expressions

- As shown in Figure D-7, notice that a cell address in a constraint expression can be a cell address in the **CHANGING CELLS** section, a cell address in the **CONSTANTS** section, a cell address in the **CALCULATIONS** section, or a cell address in the spreadsheet body.
- You'll often need to set minimum and maximum boundaries for variables. For example, the number of basketballs (MIN and MAX) varies between 30,000 and 60,000 balls.
- Often, a boundary value is zero because you want the Solver to find a non-negative result. For example, here you want only answers that yield a positive net income. You tell the Solver that the amount in the net income cell must equal or exceed zero, so the Solver does not find an answer that produces a loss.
- Machine hours must be shared between the two kinds of balls. The constraints for the shared resource are: B20 >= 39000 and B20 <= 40000, where cell B20 shows the total hours used to make both the basketballs and footballs. The shared-resource constraint seems to be the most difficult kind of constraint for students to master when learning the Solver.

Running the Solver: Mechanics

To set up the Solver, you must tell the Solver these things:

1. The cell address of the "target" variable that you are trying to maximize (or minimize, as the case may be)
2. The changing cell addresses
3. The expressions for the constraints

The Solver will put its answers in the changing cells and on a separate sheet.

Tutorial D

Beginning to Set Up the Solver

AT THE KEYBOARD

To start setting up the Solver, select Tools—Solver. The first thing you will see is a Solver Parameters window, as shown in Figure D-8. Use the Solver Parameters window to specify the target cell, the changing cells, and the constraints. (If you don't see the Solver tool under the Tools menu, you may need to activate it by going to Tools—Add-Ins and clicking the Solver Add-Ins box to install it.)

Figure D-8 Solver Parameters window

Setting the Target Cell

To set a target cell, use the following procedure:

1. The Target Cell is net income, cell B32.
2. Click in the Set Target Cell box and enter B32.
3. Max is the default; accept it here.
4. Enter a “0” for no desired net income value (Value of). DO NOT hit Enter when you finish. You’ll navigate within this window by clicking in the next input box.

Figure D-9 shows entering data in the Set Target Cell box.

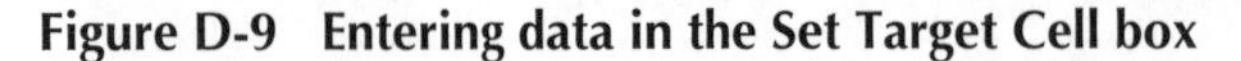

Figure D-9 Entering data in the Set Target Cell box

When you enter the cell address, Solver may put in dollar signs, as if for absolute addressing. Ignore them—do not try to delete them.

Setting the Changing Cells

The changing cells are the cells for the balls, which are in the range of cells B3:B4. Click in the By Changing Cells box and enter B3:B4, as shown in Figure D-10. (Do *not* then hit Enter.)

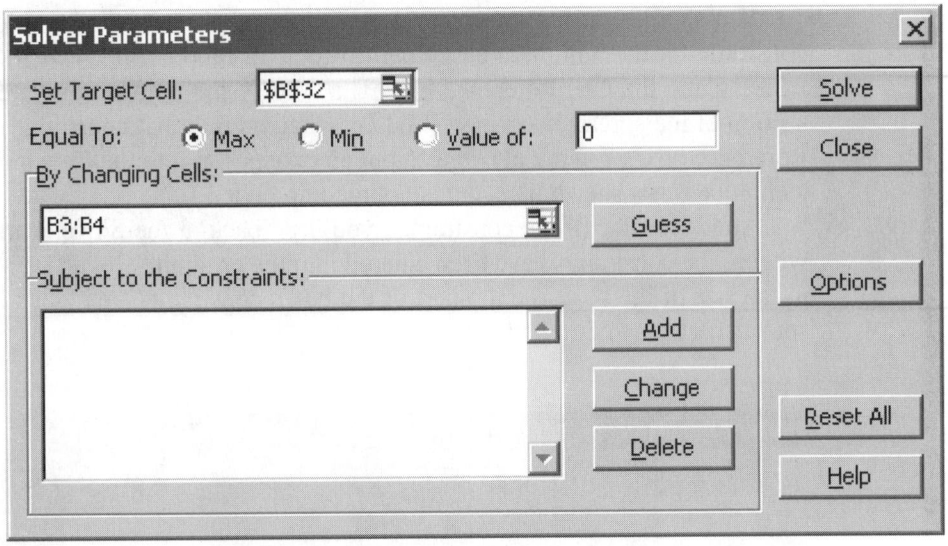

Figure D-10 Entering data in the By Changing Cells box

Entering Constraints

You are now ready to enter the constraint formulas one by one. To start, click the Add button. As shown in Figure D-11, you'll see the Add Constraint window (here, shown with the minimum basketball production constraint entered).

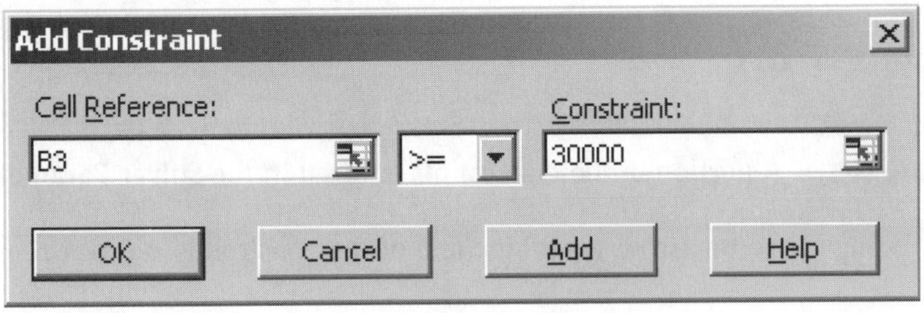

Figure D-11 Entering data in the Add Constraint window

You should note the following about entering constraints and Figure D-11:

- To enter a constraint expression, do four things: (1) Type the variable's cell address in the left Cell Reference input box; (2) select the operator (<=, =, or >=) in the smaller middle box; (3) enter the expression's right-side value, which is either a raw number or the cell address of a value, into the Constraint box; and (4) click Add to enter the constraint into the program. If you change your mind about the expression and do not want to enter it, click Cancel.
- The minimum basketballs constraint is: B3 >= 30000. Enter that constraint now. (Later, Solver may put an "equals" sign in front of the 30000 and dollar signs in the cell reference.)
- After entering the constraint formula, click the Add button. This puts the constraint into the Solver model. It also leaves you in the Add Constraint window, allowing you to enter other constraints. You should enter those now. See Figure D-7 for the logic.
- When you're done entering constraints, click the Cancel button. This takes you back to the Solver Parameters window.

You should not put an expression into the Cell Reference window. For example, the constraint for the minimum basketball-to-football ratio is B3/B4 >= 1.5. You should not put =B3/B4 into the Cell Reference box. This is why the ratio is computed in the Calculations section of the spreadsheet (in cell B17). When adding that constraint, enter B17 in the Cell Reference box. (You are allowed to put an expression into the Constraint box, although that technique is not shown here and is not recommended.)

After entering all the constraints, you'll be back at the Solver Parameters window. You will see the constraints have been entered into the program. Not all constraints will show, due to the size of the box. The top part of the box's constraints area looks like the portion of the spreadsheet shown in Figure D-12.

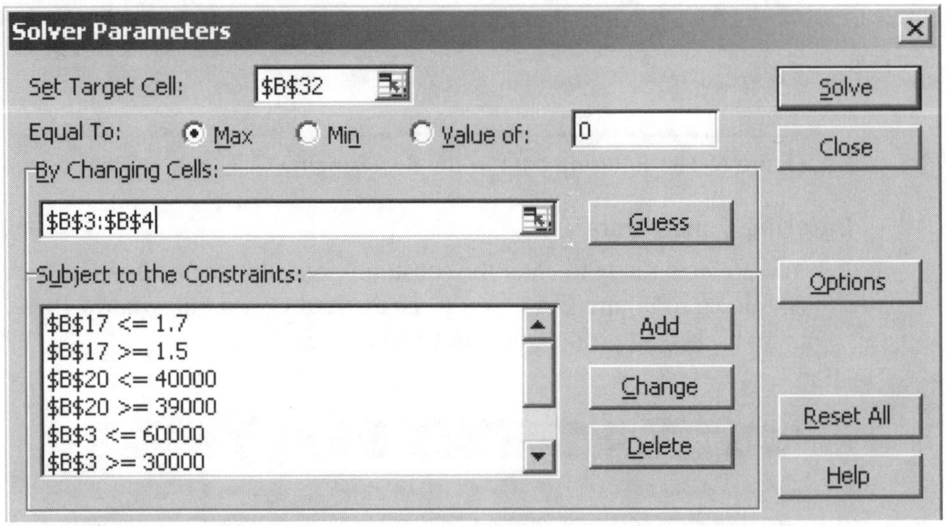

Figure D-12 A portion of the constraints entered in the Solver Parameters window

Using the scroll arrow, reveal the rest of the constraints, as shown in Figure D-13.

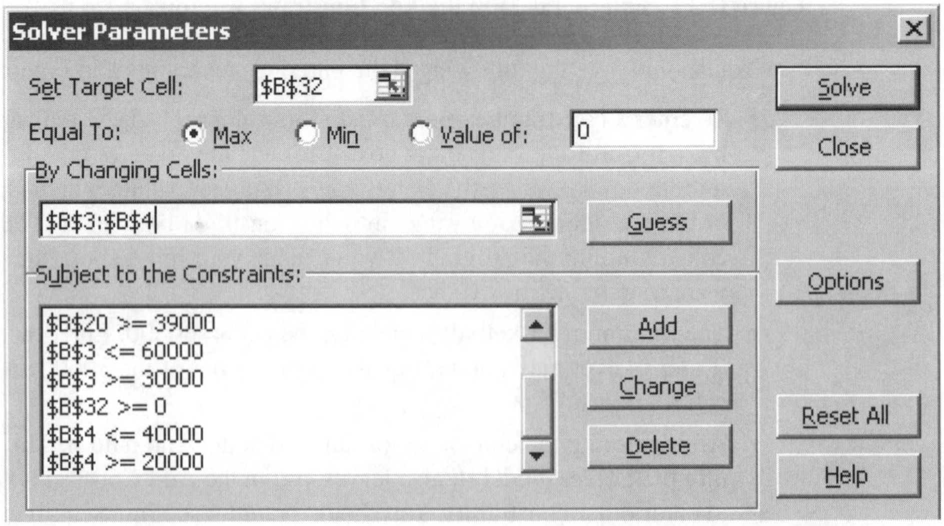

Figure D-13 Remainder of constraints entered in the Solver Parameters window

Computing the Solver's Answer

To have the Solver actually calculate answers, click Solve in the upper-right corner of the Solver Parameters window. The Solver does its work in the background—you do not see the internal calculations. Then the Solver gives you a Solver Results window, as shown in Figure D-14.

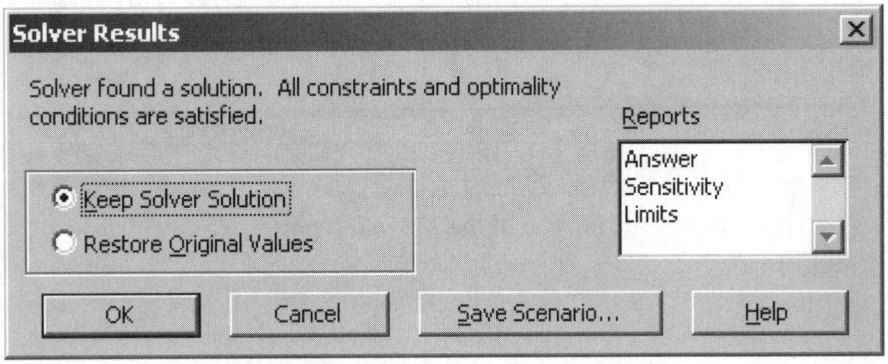

Figure D-14 Solver Results window

In the Solver Results window, the Solver tells you it has found a solution and that the optimality conditions were met. This is a very important message—you should always check for it. It means an answer was found and the constraints were satisfied.

By contrast, your constraints might be such that the Solver cannot find an answer. For example, suppose you had a constraint that said, in effect: "Net income must be at least a billion dollars." That amount cannot be reached, given so few basketballs and footballs and these prices. The Solver would report that no answer is feasible. The Solver may find an answer by ignoring some constraints. Solver would tell you that too. In either case, there would be something wrong with your model, and you would need to rework it.

There are two ways to see your answers. One way is to click OK. This lets you see the new changing cell values. A more formal (and complete) way is to click Answer in the Reports box, and then click OK. This puts detailed results into a new sheet in your Excel book. The new sheet is called an Answer Report. All answer reports are numbered sequentially as you run the Solver.

To see the Answer Report, click its tab, as shown in Figure D-15. (Here, this is Answer Report 1.)

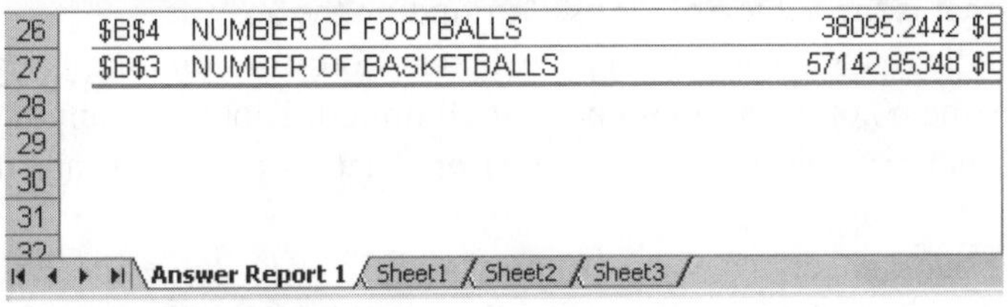

Figure D-15 Answer Report Sheet tab

This takes you to the Answer Report. The top portion of the report is shown in Figure D-16.

	A	B	C	D	E
1	Microsoft Excel 11.0 Answer Report				
2	Worksheet: [SPORTS1.XLS]Sheet1				
3					
4					
5					
6	Target Cell (Max)				
7		Cell	Name	Original Value	Final Value
8		$B$32	NET INCOME AFTER TAXES	9.90	473142.87
9					
10					
11	Adjustable Cells				
12		Cell	Name	Original Value	Final Value
13		$B$3	NUMBER OF BASKETBALLS	1	57142.85348
14		$B$4	NUMBER OF FOOTBALLS	1	38095.2442

Figure D-16 Top portion of the Answer Report

Figure D-17 shows the remainder of the Answer Report.

	A	B	C	D	E	F
17	Constraints					
18		Cell	Name	Cell Value	Formula	Status
19		$B$32	NET INCOME AFTER TAXES	473142.87	$B$32>=0	Not Binding
20		$B$17	RATIO OF BASKETBALLS TO FOOTBALLS	1.50	$B$17<=1.7	Not Binding
21		$B$17	RATIO OF BASKETBALLS TO FOOTBALLS	1.50	$B$17>=1.5	Binding
22		$B$20	TOTAL MACHINE HOURS USED (BB + FB)	40000.00	$B$20<=40000	Binding
23		$B$20	TOTAL MACHINE HOURS USED (BB + FB)	40000.00	$B$20>=39000	Not Binding
24		$B$3	NUMBER OF BASKETBALLS	57142.85348	$B$3>=30000	Not Binding
25		$B$4	NUMBER OF FOOTBALLS	38095.2442	$B$4<=40000	Not Binding
26		$B$4	NUMBER OF FOOTBALLS	38095.2442	$B$4>=20000	Not Binding
27		$B$3	NUMBER OF BASKETBALLS	57142.85348	$B$3<=60000	Not Binding

Figure D-17 Remainder of Answer Report

At the beginning of this tutorial, the changing cells had a value of 1, and the income was $9.90 (Original Value). The optimal solution values (Final Value) are also shown: $473,142.87 for net income (the target), and 57,142.85 basketballs and 38,095.24 footballs for the changing (adjustable) cells. (Of course, you cannot make a part of a ball. The Solver can be asked to find only integer solutions; this technique is discussed at the end of this tutorial.)

The report also shows detail for the constraints: the constraint expression and the value that the variable has in the optimal solution. "Binding" means the final answer caused Solver to bump up against the constraint. For example, the maximum number of machine hours was 40,000, and that is the value Solver used in finding the answer.

"Not Binding" means the reverse. A better word for "binding" might be "constraining." For example, the 60,000 maximum basketball limit did not constrain the Solver.

The procedures used to change (edit) or delete a constraint are discussed later in this tutorial.

Print the worksheets (Answer Report and Sheet1). Save the Excel file (File—Save). Then, use File—Save As to make a new file called **SPORTS2.xls**, to be used in the next section of this tutorial.

EXTENDING THE EXAMPLE

Next, you'll modify the sporting goods spreadsheet. Suppose that management wants to know what net income would be if certain constraints were changed. In other words, management wants to play "what-if" with certain Base Case constraints. The resulting second case is called the Extension Case. Let's look at some changes to the original Base Case conditions.

- Assume that maximum production constraints will be removed.
- Similarly, the basketball-to-football production ratios (1.5, 1.7) will be removed.
- There will still be minimum production constraints at some low level: Assume that at least 30,000 basketballs and 30,000 footballs will be produced.
- The machine-hours shared resource imposes the same limits as it did previously.
- A more ambitious profit goal is desired: The ratio of net income after taxes to total revenue should be greater than or equal to .33. This constraint will replace the constraint calling for profits greater than zero.

AT THE KEYBOARD

Begin by putting 1s in the changing cells. You will need to compute the ratio of net income after taxes to total revenue. Enter that formula in cell B21. (The formula should have the net income after taxes cell address in the numerator and the total revenue cell address in the denominator.) In the Extension Case, the value of this ratio for the Solver's optimal answer must be at least .33. Click the Add button and enter that constraint.

Then, in the Solver Parameters window, constraints that are no longer needed are highlighted (select by clicking) and deleted (click the Delete button). Do that for the net income >= 0 constraint, the maximum football and basketball constraints, and the basketball-to-football ratio constraints.

The minimum football constraint must be modified, not deleted. Select that constraint, then click Change. That takes you to the Add Constraint window. Edit the constraint so 30,000 is the lower boundary.

When you are finished with the constraints, your Solver Parameters window should look like the one shown in Figure D-18.

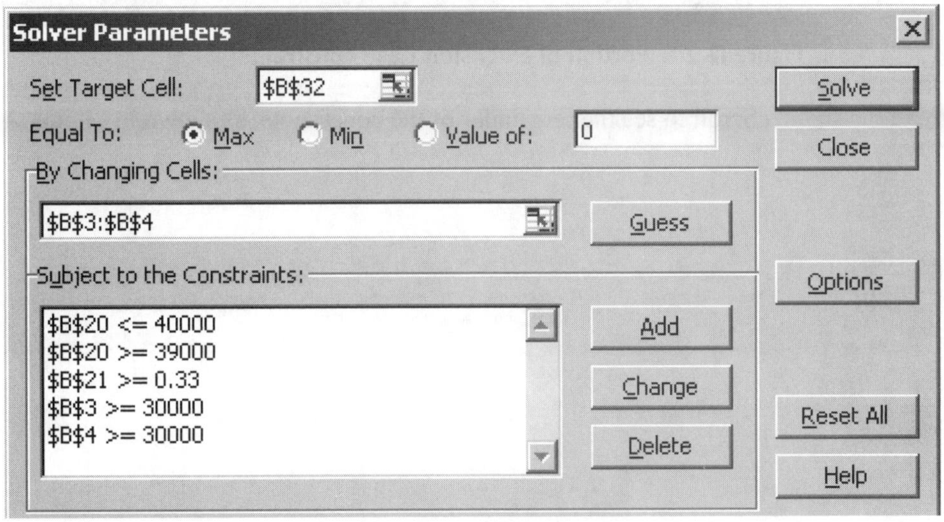

```
Solver Parameters
Set Target Cell: $B$32
Equal To: (•) Max  ( ) Min  ( ) Value of: 0
By Changing Cells:
$B$3:$B$4                                  Guess
Subject to the Constraints:
$B$20 <= 40000                             Add
$B$20 >= 39000                             Change
$B$21 >= 0.33                              Delete
$B$3 >= 30000
$B$4 >= 30000
Solve  Close  Options  Reset All  Help
```

Figure D-18 Extension Case Solver Parameters window

You can tell Solver to solve for integer values. Here, cells B3 and B4 should be whole numbers. You use the Int constraint to do that. Figure D-19 shows entering the Int constraint.

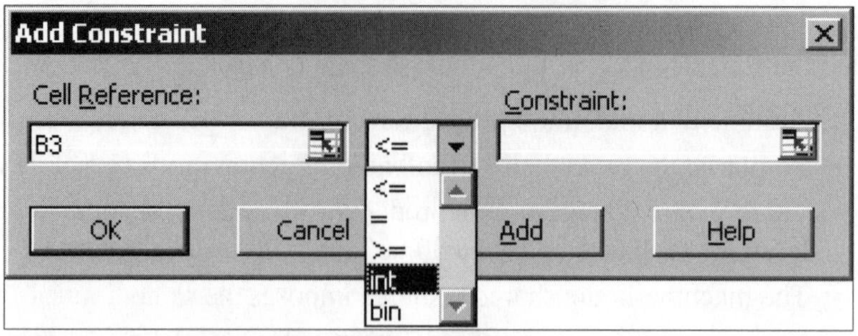

Figure D-19 Entering the Int constraint

Make those constraints for the changing cells. Your constraints should now look like the beginning portion of those shown in Figure D-20.

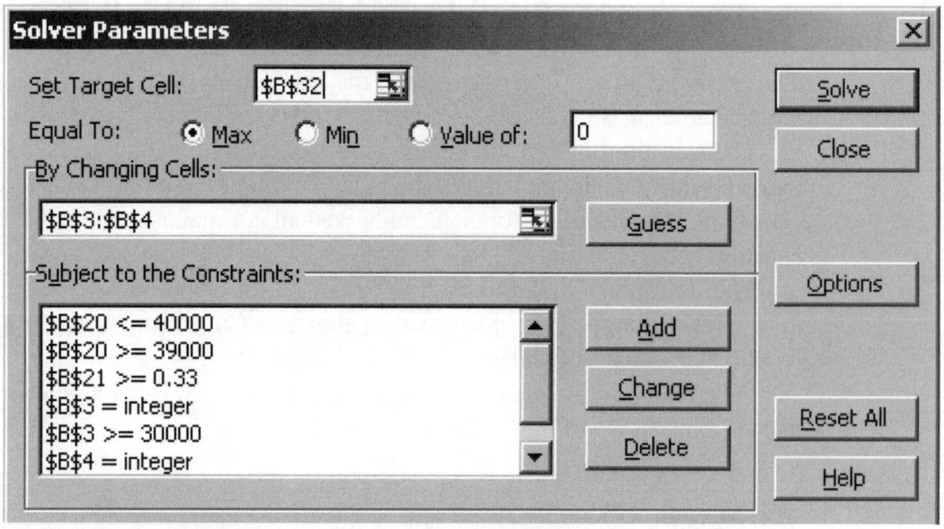

Figure D-20 Portion of Extension Case constraints

Scroll to see the remainder of the constraints, as shown in Figure D-21.

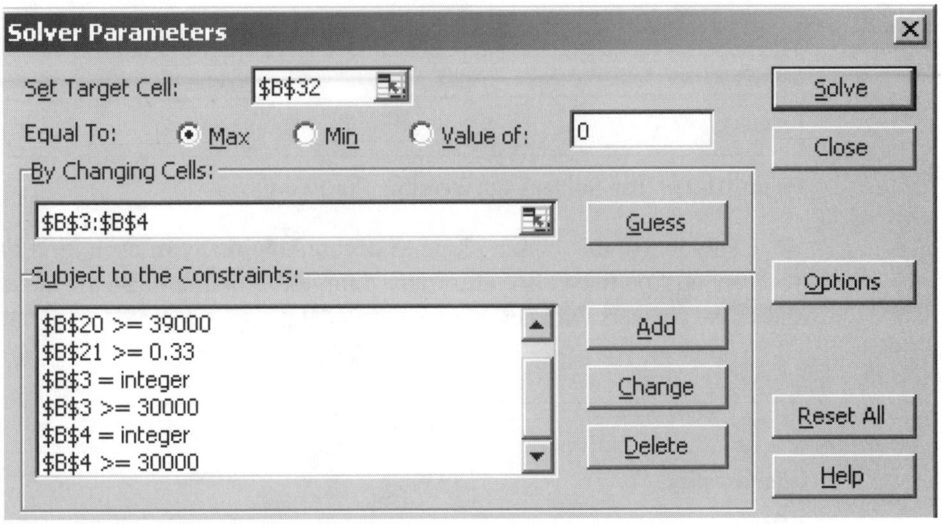

Solver Parameters

Set Target Cell: $B$32

Equal To: (•) Max ( ) Min ( ) Value of: 0

By Changing Cells:

$B$3:$B$4

Subject to the Constraints:

$B$20 >= 39000
$B$21 >= 0.33
$B$3 = integer
$B$3 >= 30000
$B$4 = integer
$B$4 >= 30000

Solve | Close | Guess | Options | Add | Change | Delete | Reset All | Help

Figure D-21 Remainder of Extension Case constraints

The constraints are now only for the minimum production levels, the ratio of net income after taxes to total revenue, machine-hours shared resource constraints, and whole number output. When the Solver is run, the values in the Answer Report look like those shown in Figure D-22.

	A	B	C	D	E	F
6	Target Cell (Max)					
7		Cell	Name	Original Value	Final Value	
8		$B$32	NET INCOME AFTER TAXES	9.90	556198.38	
9						
10						
11	Adjustable Cells					
12		Cell	Name	Original Value	Final Value	
13		$B$3	NUMBER OF BASKETBALLS	1	30000	
14		$B$4	NUMBER OF FOOTBALLS	1	83333	
15						
16						
17	Constraints					
18		Cell	Name	Cell Value	Formula	Status
19		$B$20	TOTAL MACHINE HOURS USED (BB + FB)	39999.90	$B$20<=40000	Not Binding
20		$B$20	TOTAL MACHINE HOURS USED (BB + FB)	39999.90	$B$20>=39000	Not Binding
21		$B$21	RATIO OF NET INCOME TO REVENUE	0.416109655	$B$21>=0.33	Not Binding
22		$B$3	NUMBER OF BASKETBALLS	30000	$B$3>=30000	Binding
23		$B$4	NUMBER OF FOOTBALLS	83333	$B$4>=30000	Not Binding
24		$B$3	NUMBER OF BASKETBALLS	30000	$B$3=integer	Binding
25		$B$4	NUMBER OF FOOTBALLS	83333	$B$4=integer	Binding

Figure D-22 Extension Case Answer Report

The Extension Case answer differs from the Base Case answer. Which production schedule should management use? The one that has maximum production limits? Or the one that has no such limits? These questions are posed to get you to think about the purpose of using a DSS program. Two scenarios, the Base Case and the Extension Case, were modeled in the Solver. The very different answers are shown in Figure D-23.

	Base Case	Extension Case
Basketballs	57,143	30,000
Footballs	38,095	83,333

Figure D-23 The Solver's answers for the two cases

Can you use this output alone to decide how many of each kind of ball to produce? No, you cannot. You must also refer to the "Target," which in this case is net income. Figure D-24 shows the answers with net income target data.

	Base Case	Extension Case
Basketballs	57,143	30,000
Footballs	38,095	83,333
Net Income	$473,143	$556,198

Figure D-24 The Solver's answers for the two cases—with target data

Viewed this way, the Extension Case production schedule looks better, because it gives you a higher target net income.

At this point, you should save the **SPORTS2.xls** file (File—Save) and then close it (File—Close).

USING THE SOLVER ON A NEW PROBLEM

Here is a short problem that will let you test what you have learned about the Excel Solver.

Setting Up the Spreadsheet

Assume that you run a shirt-manufacturing company. You have two products: (1) polo-style T-shirts, and (2) dress shirts with button-down collars. You must decide how many T-shirts and how many button-down shirts to make. Assume that you'll sell every shirt you make.

AT THE KEYBOARD

Open a file called **SHIRTS.xls**. Set up a Solver spreadsheet to handle this problem.

CHANGING CELLS Section

Your changing cells should look like those shown in Figure D-25.

	A	B
1	SHIRT MANUFACTURING EXAMPLE	
2	CHANGING CELLS	
3	NUMBER OF T-SHIRTS	1
4	NUMBER OF BUTTON-DOWN SHIRTS	1

Figure D-25 Shirt manufacturing changing cells

CONSTANTS Section

Your spreadsheet should contain the constants shown in Figure D-26. A discussion of constant cells (and some of your company's operations) follows the figure.

	A	B
6	**CONSTANTS**	
7	TAX RATE	0.28
8	SELLING PRICE: T-SHIRT	8.00
9	SELLING PRICE: BUTTON-DOWN SHIRT	36.00
10	VARIABLE COST TO MAKE: T-SHIRT	2.50
11	VARIABLE COST TO MAKE: BUTTON-DOWN SHIRT	14.00
12	COTTON USAGE (LBS): T-SHIRT	1.50
13	COTTON USAGE (LBS): BUTTON-DOWN SHIRT	2.50
14	TOTAL COTTON AVAILABLE (LBS)	13000000
15	BUTTONS PER T-SHIRT	3.00
16	BUTTONS PER BUTTON-DOWN SHIRT	12.00
17	TOTAL BUTTONS AVAILABLE	110000000

Figure D-26 Shirt manufacturing constants

- The TAX RATE is .28 on pre-tax income, but no taxes are paid on losses.
- SELLING PRICE: You sell polo-style T-shirts for $8 and button-down shirts for $36.
- VARIABLE COST TO MAKE: It costs $2.50 to make a T-shirt and $14 to make a button-down shirt. These variable costs are for machine-operator labor, cloth, buttons, and so forth.
- COTTON USAGE: Each polo T-shirt uses 1.5 pounds of cotton fabric. Each button-down shirt uses 2.5 pounds of cotton fabric.
- TOTAL COTTON AVAILABLE: You have only 13 million pounds of cotton on hand to be used to make all the T-shirts and button-down shirts.
- BUTTONS: Each polo T-shirt has 3 buttons. By contrast, each button-down shirt has 1 button on each collar tip, 8 buttons down the front, and 1 button on each cuff, for a total of 12 buttons. You have 110 million buttons on hand to be used to make all your shirts.

CALCULATIONS Section

Your spreadsheet should contain the calculations shown in Figure D-27.

	A	B
19	**CALCULATIONS**	
20	RATIO OF NET INCOME TO TOTAL REVENUE	
21	COTTON USED: T-SHIRTS	
22	COTTON USED: BUTTON-DOWN SHIRTS	
23	COTTON USED: TOTAL	
24	BUTTONS USED: T-SHIRTS	
25	BUTTONS USED: BUTTON-DOWN SHIRTS	
26	BUTTONS USED: TOTAL	
27	RATIO OF BUTTON-DOWNS TO T-SHIRTS	

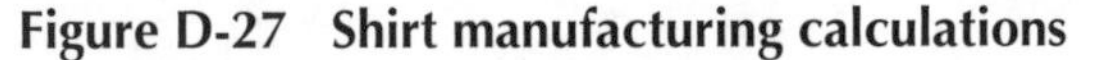

Figure D-27 Shirt manufacturing calculations

Calculations (and related business constraints) are discussed next.

- RATIO OF NET INCOME TO TOTAL REVENUE: The minimum return on sales (ratio of net income after taxes divided by total revenue) is .20.
- COTTON USED/BUTTONS USED: You have a limited amount of cotton and buttons. The usage of each resource must be calculated, then used in constraints.
- RATIO OF BUTTON-DOWNS TO T-SHIRTS: You think you must make at least 2 million T-shirts and at least 2 million button-down shirts. You want to be known as a balanced shirtmaker, so you think that the ratio of button-downs to T-shirts should be no greater than 4:1. (Thus, if 9 million button-down shirts and 2 million T-shirts were produced, the ratio would be too high.)

INCOME STATEMENT Section

Your spreadsheet should have the income statement skeleton shown in Figure D-28.

	A	B
29	**INCOME STATEMENT**	
30	T-SHIRT REVENUE	
31	BUTTON-DOWN SHIRT REVENUE	
32	TOTAL REVENUE	
33	VARIABLE COSTS: T-SHIRTS	
34	VARIABLE COSTS: BUTTON-DOWNS	
35	TOTAL COSTS	
36	INCOME BEFORE TAXES	
37	INCOME TAX EXPENSE	
38	NET INCOME AFTER TAXES	

Figure D-28 Shirt manufacturing income statement line items

The Solver's target is net income, which must be maximized.

Use the table shown in Figure D-29 to write out your constraints before entering them into the Solver.

Expression in English	Fill in the Excel Expression
Net income to revenue	_____ >= _____
Ratio of BDs to Ts	_____ <= _____
Min T-shirts	_____ >= _____
Min button-downs	_____ >= _____
Usage of buttons	_____ <= _____
Usage of cotton	_____ <= _____

Figure D-29 Logic of shirt manufacturing constraints

When you are finished with the program, print the sheets. Then, use File—Save, File—Close, and then File—Exit, to leave Excel.

TROUBLE-SHOOTING THE SOLVER

Use this section to overcome problems with the Solver and as a review of some Windows file-handling procedures.

Rerunning a Solver Model

Assume that you have changed your spreadsheet in some way and want to rerun the Solver to get a new set of answers. (For example, you may have changed a constraint or a formula in your spreadsheet.) Before you click Solve again to rerun the Solver, you should put the number 1 in the changing cells. The Solver can sometimes give odd answers if its point of departure is a set of prior answers.

Creating Over-Constrained Models

It is possible to set up a model that has no logical solution. For example, in the second version of the sporting goods problem, suppose that you had specified that at least 1 million basketballs were needed. When you clicked Solve, the Solver would have tried to compute an answer, but then would have admitted defeat by telling you that no feasible solution is possible, as shown in Figure D-30.

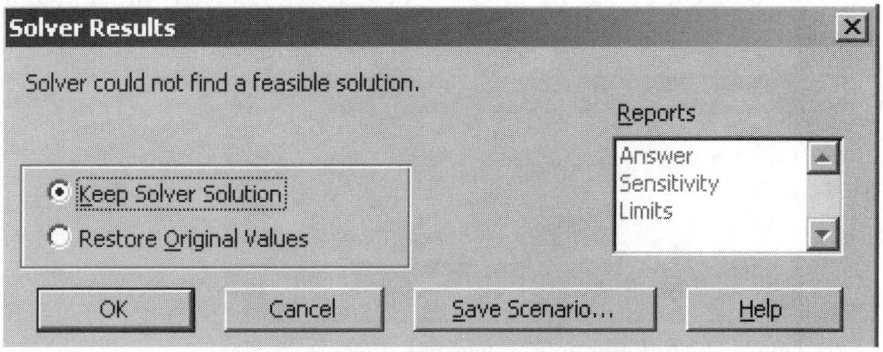

Solver Results

Solver could not find a feasible solution.

Reports

Answer
Sensitivity
Limits

Keep Solver Solution
Restore Original Values

OK | Cancel | Save Scenario... | Help

Figure D-30 Solver Results message: Solution not feasible

In the Reports window, the choices (Answer, etc.) would be in gray—indicating they are not available as options. Such a model is sometimes called "over-constrained."

Setting a Constraint to a Single Amount

It's possible you'll want an amount to be a specific number, as opposed to a number in a range. For example, if the number of basketballs needed to be exactly 30,000, then the "equals" operator would be selected, as shown in Figure D-31.

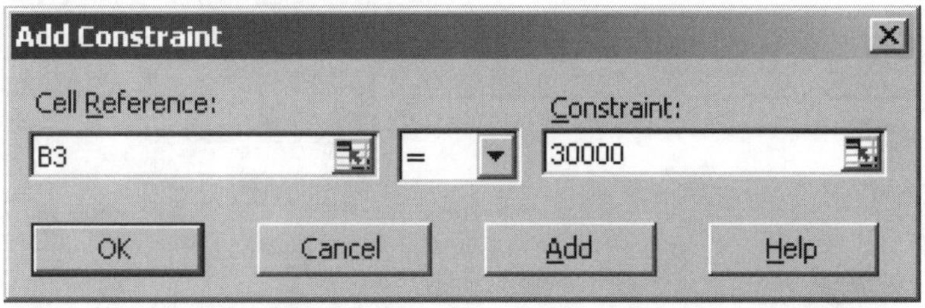

Add Constraint

Cell Reference: B3 | = | Constraint: 30000

OK | Cancel | Add | Help

Figure D-31 Constraining a value to equal a specific amount

Setting a Changing Cell to an Integer

You may want to force changing cell values to be integers. The way to do that is to select the Int operator in the Add Constraint window. This was described in a prior section.

Forcing the Solver to find only integer solutions slows down the Solver. In some cases, the change in speed can be noticeable to the user. Doing this can also prevent the Solver from seeing a feasible solution—when one can be found if the Solver is allowed to find non-integer answers. For these reasons, it's usually best not to impose the integer constraint unless the logic of the problem demands it.

Deleting Extra Answer Sheets

Suppose that you've run different scenarios, each time asking for an Answer Report. As a result, you have a number of Answer Report sheets in your Excel file, but you don't want to keep them all. How do you get rid of an Answer Report sheet? Follow this procedure: First, get the unwanted Answer Report sheet on the screen by clicking the sheet's tab. Then select Edit—Delete Sheet. You will be asked if you really mean it. If you do, click accordingly.

Restarting the Solver with All-New Constraints

Suppose that you wanted to start over, with a new set of constraints. In the Solver Parameters window, click Reset All. You will be asked if you really mean it, as shown in Figure D-32.

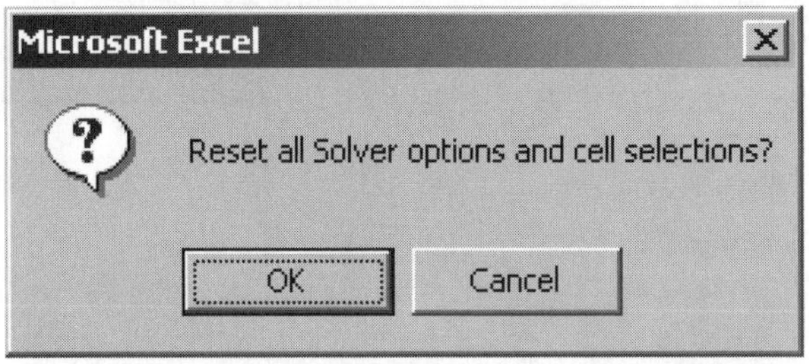

Figure D-32 Reset options warning query

If you do, then select OK. This gives you a clean slate, with all entries deleted, as shown in Figure D-33.

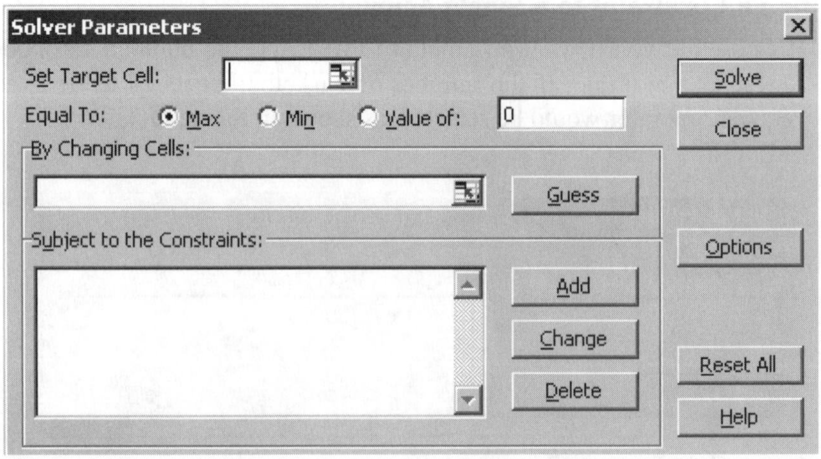

Figure D-33 Reset Solver Parameters window

As you can see, the target cell, changing cells, and constraints have been reset. From this point, you can specify a new model.

If you select Reset All, you really are starting over. If you merely want to add, delete, or edit a constraint, do not use Reset All. Use the Add, Delete, or Change buttons, as the case may be.

Solver Options Window

The Solver has a number of internal settings that govern its search for an optimal answer. If you click the Options button in the Solver Parameters window, you will see the defaults for these settings, as shown in Figure D-34:

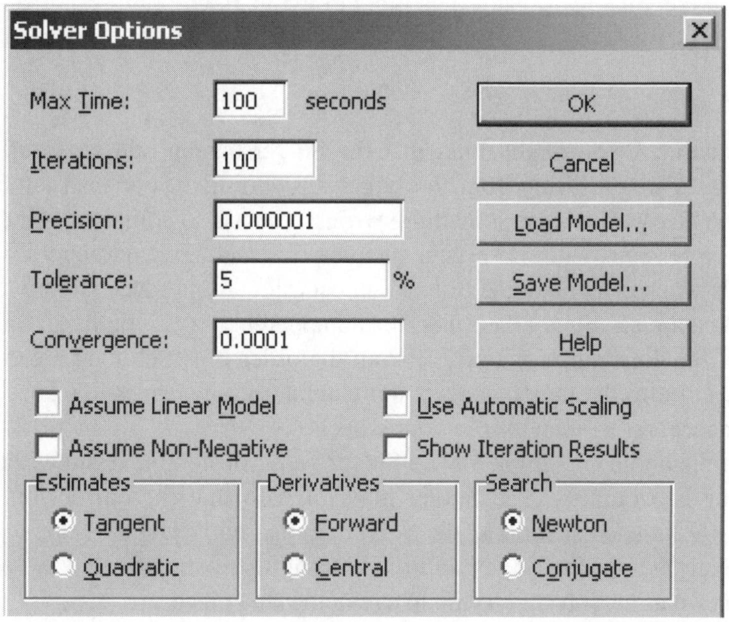

Figure D-34 Solver Options window, with default settings for Solver Parameters

Very broadly speaking, Solver Options govern how long the Solver works on a problem and/or how precise it must be in satisfying constraints. You should not check Assume Linear Model if changing cells are multiplied or divided (as they are in this book's cases) or if some of the spreadsheet's formulas use exponents.

You should not need to change these default settings for the cases in this book. If you think that your Solver work is correct, but Solver cannot find a feasible solution, you should check to see that Solver Options are set as shown in Figure D-34.

Printing Cell Formulas in Excel

To show the Cell Formulas on the screen, press the Ctrl and the left quote (`) keys at the same time: Ctrl-`. (The left quote is usually on the same key as the tilde [~].) This automatically widens cells so the formulas can be read. You can change cell widths by clicking and dragging at the column indicator (A, B, C…) boundaries. Another way to show the cells' formulas is to use Tools—Formula Auditing—Formula Auditing Mode.

To print the formulas, just use File—Print. Print the sheet as you would normally. To restore the screen to its typical appearance (showing values, not formulas), press Ctrl-` again. (It's like a toggle switch.) If you did not change any column widths when in the cell formula view, the widths will be as they were.

Review of Printing, Saving, and Exiting Procedures

Print the Solver spreadsheets in the normal way. Activate the sheet, then select File—Print. You can print an Answer Report sheet in the same way.

To save a file, use File—Save, or File—Save As. Be sure to select Drive A: in the Drive window, if you intend your file to be on a disk. When exiting from Excel, always start with File—Close (with the disk in Drive A:), then select File—Exit. Only then should you take the disk out of Drive A:.

If you merely use File—Exit (not closing first), you risk losing your work.

Sometimes, you might think that the Solver has an odd sense of humor. For instance, your results might differ from the target answers that your instructor provides for a case. Thinking that you've done something wrong, you ask to compare your cell formulas and constraint expressions with those your instructor created. Lo and behold, you can see no differences! It is surprising that the Solver can occasionally produce slightly different outputs from inputs that are seemingly the same, for no apparent reason. Perhaps, for your application, the order of the constraints matters, or even the order in which they are entered. In any case, if you are close to the target answers but cannot see any errors, it's best to see your instructor for guidance, rather than to spin your wheels.

Here is another example of the Solver sense of humor. Assume that you ask for Integer changing cell outputs. The Solver may tell you that the correct output is 8.0000001, or 7.9999999. In such situations, the Solver is apparently just not sure about its own rounding! You merely humor the Solver and (continuing the example) take the result as the integer 8, which is what the Solver is trying to say in the first place.

The Public TV Fundraising Decision

DECISION SUPPORT USING EXCEL

PREVIEW

In this case, you will use the Excel Solver to help a public television station decide what promotional items to offer in its annual fundraising event. The goal is to raise the most money for the least cost.

PREPARATION

- Review spreadsheet concepts discussed in class and/or in your textbook.
- Complete any exercises that your instructor assigns.
- Complete any part of Tutorial D that your instructor assigns, or refer to it as necessary.
- Review file-saving procedures for Windows programs. These are discussed in Tutorial C.
- Refer to Tutorial E as necessary.

BACKGROUND

WHCZ is the public television station that serves your area. Public TV stations have few advertisers and often have trouble making ends meet. Like most public TV stations, once a year WHCZ has a week-long campaign in which they ask the public to donate money to support the station's programming.

During the campaign, the station devotes some of each programming hour to beseeching viewers to send in money. To induce people to donate, the station gives away promotional "gifts," such as fancy coffee cups or gift certificates to local restaurants, to those who donate at certain times. This year, however, the station will offer different gifts. You have been called in to help the station decide how to use the promotional gifts.

Two promotional gift items are under consideration: (1) large blue-and-white golf umbrellas with the WHCZ logo, and (2) one-year subscriptions to *NewsTime*, a popular weekly news magazine. If a viewer gives $40, he or she will get the umbrella, which represents a substantial discount from market value. If a viewer gives $50, he or she will get a year's subscription to the magazine, again at a substantial discount from market value.

WHCZ's campaign will last 7 days, for 12 hours a day. The station will do one "sales pitch" each hour and will offer either the umbrella or the magazine, but not both in the same sales pitch. Each sales pitch can be considered an advertising "spot."

Unfortunately, the station must commit to buying the promotional items beforehand. The station will pay $20 for each umbrella and $25 for each magazine subscription. The station cannot store leftover umbrellas and must give away every item it buys.

The question for management is this: How many magazine spots and how many umbrella spots should the station air during the entire campaign? The goal is, of course, to maximize income. Management turns to you for help, knowing that you have studied the Excel Solver for this sort of optimization decision.

Here are some other data about the situation:

- The typical audience size during any hour of the campaign is expected to be 100,000 homes. Of course, the audience will not be the same 100,000 homes every hour. Marketing experts have been consulted about buying habits in the area. They say that 1% of the homeowners will respond positively to an umbrella spot, and that 1.5% will respond positively to a magazine spot. For example, if 10 umbrella spots are run, 100,000 decision-makers see each spot, and the result is 10 * 100,000 * .01 positive umbrella responses.
- Based on prior campaign history, station management knows that some viewers who make a donation for a gift will spontaneously donate money again, later in the year. Viewers who make a second donation give an average of $20. Management estimates that 25% of umbrella recipients and 15% of magazine recipients will be repeat donors.
- Management thinks that some balance in the spots is needed. They think that at least 30 of each kind of spot should be run. Of course, they cannot run part of an umbrella spot or part of a magazine spot.
- Management wants to earn as much net income as possible in the campaign. The ratio of net income to total revenue should be at least 50%.

Management wants an answer to this question: How many of each spot should be run in the campaign? Knowing this will let management compute how many umbrellas and magazine subscriptions should be ordered for the campaign.

You must model two situations: a Base Case (the parameters of which have just been described) and an alternative Extension Case (the parameters of which will be given shortly).

You will use the Solver as your decision-support modeling tool. After you have finished the Base Case, you will modify its spreadsheet to create the Extension Case, which will let you play "what if" with the decision. Station management will use your results to decide how to structure the promotional campaign. Finally, you will write a memorandum that describes the problem and recommends a course of action. Your memorandum will include (as attachments) printouts of the Solver spreadsheets and reports documenting each model. You will give your instructor your memo, printouts, and your disk or CD.

ASSIGNMENT 1 CREATING A SPREADSHEET FOR DECISION SUPPORT

In this assignment, you will produce a spreadsheet that models the business decision. In Assignment 1A, you will make a Solver spreadsheet to model the Base Case. In Assignment 1B, you will make a Solver spreadsheet to model the Extension Case. In Assignment 2, you will use the spreadsheet models to develop information needed to recommend the best promotional product mix, and then you will document your recommendation in a memorandum to station management. In Assignment 3, you will give your recommendation in an oral presentation.

Next, you will create the spreadsheet models of the promotion-mix decision. Your spreadsheet should have the following sections:

- CHANGING CELLS
- CONSTANTS
- CALCULATIONS
- INCOME STATEMENT

Your spreadsheets will also include the decision constraints.

A Base Case spreadsheet skeleton is available to you, so you need not type in the skeleton. *To access the spreadsheet skeleton, go to your Data files, select Case 8, and then select **WHCZ1.xls**.*

Assignment 1A: Creating the Spreadsheet—Base Case

You will model the promotion problem described. Your model, when run, will tell management how many of each kind of spot to run.

A discussion of each spreadsheet section follows. The discussion is about: (1) how each section should be set up, and (2) the logic of the sections' cell formulas.

CHANGING CELLS Section

Your spreadsheet should have the changing cells shown in Figure 8-1.

	A	B	C	D
1	WHCZ TV PROMOTION DECISION -- BASE CASE			
2	CHANGING CELLS			
3	NUMBER OF UMBRELLA SPOTS		1	
4	NUMBER OF NEWS MAGAZINE SPOTS		1	
5				

Figure 8-1 CHANGING CELLS section

You are asking the Solver to compute the number of each type of spot. Start with a 1 in each cell. The Solver will change each 1 as it computes the answer. Do not allow the Solver to recommend a fraction of a spot.

CONSTANTS Section

Your spreadsheet should have the constants shown in Figure 8-2. An explanation of the line items follows the figure.

	A	B	C
6	**CONSTANTS**		
7	CONTRIBUTION FOR UMBRELLA		40
8	CONTRIBUTION FOR MAGAZINE		50
9	COST OF UMBRELLA		20
10	COST OF MAGAZINE		25
11	REPEAT REVENUE DONATION		20
12	REPEAT REVENUE % -- UMBRELLA		0.250
13	REPEAT REVENUE % -- MAGAZINE		0.150
14	NUMBER OF DAYS IN CAMPAIGN		7
15	NUMBER OF HOURS IN CAMPAIGN DAY		12
16	AVERAGE AUDIENCE SIZE (HOMES)		100000
17	AUDIENCE RESPONSE % -- UMBRELLA		0.010
18	AUDIENCE RESPONSE % -- MAGAZINE		0.015

Figure 8-2 CONSTANTS section

- CONTRIBUTION FOR UMBRELLA: Umbrella spots bring in $40 per contributor.
- CONTRIBUTION FOR MAGAZINE: Magazine spots bring in $50 per contributor.
- COST OF UMBRELLA: An umbrella costs the station $20.
- COST OF MAGAZINE: A magazine subscription costs the station $25.
- REPEAT REVENUE DONATION: Repeat contributors will give $20 on average.
- REPEAT REVENUE % — UMBRELLA: 25% of those who donate to get an umbrella will later donate again.
- REPEAT REVENUE % — MAGAZINE: 15% of those who donate to get a magazine subscription will later donate again.
- NUMBER OF DAYS IN CAMPAIGN: The campaign will last 7 days.
- NUMBER OF HOURS IN CAMPAIGN DAY: A campaign day will have spots for 12 hours, one spot per hour.
- AVERAGE AUDIENCE SIZE (HOMES): On average, 100,000 homes will be tuned to WHCZ during the spots.
- AUDIENCE RESPONSE % — UMBRELLA: 1% of those who see an umbrella spot will opt to contribute.
- AUDIENCE RESPONSE % — MAGAZINE: 1.5% of those who see a magazine spot will opt to contribute.

CALCULATIONS Section

Your spreadsheet should calculate the amounts shown in Figure 8-3. The amounts will be used in the INCOME STATEMENT section or in the CONSTRAINTS. Calculations are based on values in the CHANGING CELLS section and/or in the CONSTANTS section and/or on other calculated values. An explanation of the line items follows the figure.

	A	B	C
20	**CALCULATIONS**		
21	ALLOWED NUMBER OF SPOTS		
22	CONTRIBUTORS -- UMBRELLA		
23	CONTRIBUTORS -- MAGAZINE		
24	CONTRIBUTIONS -- UMBRELLA		
25	CONTRIBUTIONS -- MAGAZINE		
26	UMBRELLA COSTS		
27	MAGAZINE COSTS		
28	REPEAT REVENUE -- UMBRELLA		
29	REPEAT REVENUE -- MAGAZINE		
30	TOTAL NUMBER OF SPOTS		
31	RATIO OF INCOME TO TOTAL REVENUE		

Figure 8-3 CALCULATIONS section

- ALLOWED NUMBER OF SPOTS: This is the maximum number of spots that can be shown in the campaign period.
- CONTRIBUTORS — UMBRELLA: This is the number of viewers that opt to contribute to get an umbrella.
- CONTRIBUTORS — MAGAZINE: This is the number of viewers that opt to contribute to get a magazine subscription.
- CONTRIBUTIONS — UMBRELLA: This is the dollar amount of revenue contributed by those who opt for the umbrella.
- CONTRIBUTIONS — MAGAZINE: This is the dollar amount of revenue contributed by those who opt for the subscription.
- UMBRELLA COSTS: This is the amount that the station spends to buy umbrellas.
- MAGAZINE COSTS: This is the amount that the station spends to buy magazine subscriptions.
- REPEAT REVENUE — UMBRELLA: This is the amount of repeat donations from those who opted for the umbrella.
- REPEAT REVENUE — MAGAZINE: This is the amount of repeat donations from those who opted for the subscription.
- TOTAL NUMBER OF SPOTS: This is the number of umbrella and magazine spots actually scheduled.
- RATIO OF INCOME TO TOTAL REVENUE: This is the ratio of net income to total revenue.

INCOME STATEMENT Section

Compute the campaign's net income, as shown in Figure 8-4. Line items are discussed after the figure.

	A	B	C
33	**INCOME STATEMENT**		
34	CONTRIBUTION REVENUE -- UMBRELLA		
35	CONTRIBUTION REVENUE -- MAGAZINE		
36	REPEAT DONATION REVENUE		
37	TOTAL REVENUE		
38	UMBRELLA COSTS		
39	MAGAZINE COSTS		
40	TOTAL COSTS		
41	NET INCOME		

Figure 8-4 INCOME STATEMENT section

- CONTRIBUTION REVENUE — UMBRELLA / MAGAZINE: This is the total revenue from magazine and umbrella contributions.
- REPEAT DONATION REVENUE: This is the total of repeat donations.
- TOTAL REVENUE: This is the total of contribution revenue and repeat donation revenue.
- UMBRELLA COSTS: This is the cost of umbrellas, a calculation that can be echoed here.
- MAGAZINE COSTS: This is the cost of magazines, a calculation that can be echoed here.
- TOTAL COSTS: This is the total of umbrella and magazine costs.
- NET INCOME: This is total revenue less total costs.

Constraints and Running the Solver

Determine the constraints. Enter the Base Case decision constraints, using the Solver. Run the Solver. Make an Answer Report when the Solver says that a solution has been found that satisfies the constraints.

When you are finished, print the entire workbook, including the Answer Report. Save the Base Case Solver spreadsheet file (File—Save; **WHCZ1.xls** should be the filename). Then, to prepare for the Extension Case, use File—SaveAs to make a new spreadsheet. (**WHCZ2.xls** should be the filename.)

Assignment 1B: Creating the Spreadsheet—Extension Case

A very wealthy local industrialist thinks that the editorial stance of the *NewsTime* magazine is unAmerican and anti-intellectual. The industrialist is a presence in the community and is on WHCZ's board of directors and is, therefore, an influential person at the station. The industrialist would prefer that the station not give away the magazine, if the decision can be economically justified. Furthermore, she thinks that the umbrellas are really good-looking and an excellent ongoing advertisement for the station, and she would like to see more of them in public. She has an alternative promotional idea, which station management has agreed to consider.

In the alternative promotion, the magazine subscription would not be offered. Only the umbrella would be offered, at $35, not $40. The industrialist will anonymously contribute $50,000 herself. Any later spontaneous contributions will be matched dollar-for-dollar by the industrialist.

Station management agrees to consider this idea and go with it if seems financially superior to the Base Case. Management would still want to achieve a 50% net income to total revenue ratio. The industrialist says that she will not stand in the way of the Base Case proposal, if that idea looks better financially.

Modify the Extension Case spreadsheet (**WHCZ2.xls**) and related constraints to reflect the industrialist's proposal. Run the Solver. Ask for an Answer Report when the Solver says that a solution has been found that satisfies the constraints. When you are done, print the entire workbook, including the Solver Answer Report. Save the worksheet when you are finished. Close the file and exit Excel.

ASSIGNMENT 2 USING THE SPREADSHEET FOR DECISION SUPPORT

You have built the Base Case and the Extension Case models because you want to know which promotional campaign yields the highest net income after taxes. You will now complete your work by: (1) using the worksheets and Answer Reports to gather the data needed to decide which promotional campaign to run, and (2) documenting your recommendation in a memorandum and (if your instructor specifies) an oral presentation.

Assignment 2A: Using the Spreadsheet to Gather Data

You have printed the spreadsheet and Answer Report sheet for each case. You can see the results for each campaign. You should summarize the key data in a table that will be included in your memorandum. The form of that table is shown in Figure 8-5.

	Base Case	Extension Case
Number of Umbrella spots		
Number of Magazine spots		Zero
Net income		

Figure 8-5 Format of table in memorandum

Assignment 2B: Documenting Your Recommendation in a Memorandum

Use MS Word to write a brief memorandum to WHCZ's station manager about the results of your analysis. Observe the following requirements:

1. Your memorandum should have a proper heading (DATE / TO / FROM / SUBJECT). You might want to use a Word memo template (**File**, click **New**, click **On my computer** in the Templates section, click the **Memos** tab, choose **Contemporary Memo**, and then click **OK**).
2. Briefly state the problem and the decision to be made, but do not provide background—you can assume the station manager is aware of the problem. You should briefly state your analytical method and state the results of the analysis. Give the station manager your recommendation, which should be keyed to the highest net income.
3. Support the recommendation graphically, by including a summary table in your Word memo, as shown in Figure 8-5. Make the table by following the procedure described next.

   Enter a table into the Word memorandum, using the following procedure:

1. Select the **Table** menu option, click **Insert**, and then click **Table**.
2. Enter the number of rows and columns.
3. Select **AutoFormat** and choose **TableGrid1**.
4. Select **OK**, and then select **OK** again.
5. Once the table is in place, merely enter data directly into its cells.

ASSIGNMENT 3 GIVING AN ORAL PRESENTATION

Your instructor might request that you also present your analysis and recommendations in an oral presentation. If so, assume that the station's manager is very impressed by your analysis, and that he asks you to give a presentation to the station's management team and Board of Directors. Prepare to explain your analysis and recommendation to the group in 10 minutes or fewer. Use visual aids or handouts that you think are appropriate. Tutorial E has guidance on how to prepare and give an oral presentation.

DELIVERABLES

Assemble the following deliverables for your instructor:

1. Printout of your memorandum
2. Spreadsheet printouts
3. Disk or CD, which should have your Word memorandum and Excel spreadsheet files

Staple the printouts together, with the memorandum on top. Hand-write your instructor a note, stating the names of your Base Case and Extension Case *.xls* files.

The Condominium Development Decision

Decision Support Using Excel

Preview

In this case, you will use the Excel Solver to help a real estate developer decide whether a condominium project can be profitable and, if so, to prepare for negotiations with the Town Council, who regulate real estate development.

Preparation

- Review spreadsheet concepts discussed in class and/or in your textbook.
- Complete any exercises that your instructor assigns.
- Complete any part of Tutorial D that your instructor assigns, or refer to it as necessary.
- Review file-saving procedures for Windows programs. These are discussed in Tutorial C.
- Refer to Tutorial E as necessary.

Background

An Eastern college town has a sprawling bar called The Prisoner's Dilemma. Legend has it that General Washington's forces held British troops there during the Revolutionary War, which explains the first part of the bar's name. (No one is quite sure where the second part of the name comes from.) The fact that the building was not constructed until during America's *Civil* War has done nothing to refute the legend, which is passed on from one college class to the next.

The bar occupies a large lot in the middle of town. The bar has served as a raucous hangout for many generations of college students. However, the family who owns the property wants to get out of the bar business, and they've sold the property to Blue Spruce Development Company. Blue Spruce wants to tear down the bar and put up a multistory condominium building.

Everyone agrees that there is a great need for apartment and commercial space in the town. Blue Spruce CEO, Josh "Nails" Harrelson, thinks that a 70–75 unit condominium building can be put on the acre and a half that the bar occupies, along with ground floor retail and office space and underground parking. Harrelson thinks that the condos can be priced reasonably and still leave a healthy profit for Blue Spruce.

In this town, all building proposals must be approved by the Town Council. Harrelson thinks that most of the council members have an anti-development bias, and he expects some resistance to his proposed project. He expects this objection: The Prisoner's Dilemma has no parking—everyone finds their way to it on foot. The proposed building would add traffic and create a need for parking. Nails Harrelson knows that the Town Council will make a big issue out of the added traffic congestion and parking problems. In fact, some council members have informally advised him that they would want to cut back the scope of the condominium project. Harrelson is not sure whether he can make money if the number of units is drastically reduced.

Harrelson turns to you to find the answers to two questions:

1. Will he make an acceptable return (by his standards) if he builds the condominium that he wants to build?
2. If he builds the condominium that some council members are rumored to want, can he make any money at all?

Harrelson knows that you have studied the Excel Solver and can help him with this sort of optimization decision. Harrelson's goal is, of course, to maximize after-tax net income from construction. Data about the condominium that Harrelson wants to build is set forth next.

The Construction Site

The plot of ground is square in shape and occupies about 1.5 acre. Each side of the square has 250 feet of usable space. The ground sits between the college town's two main roads, Main Street and Oak Street, one-way streets running in opposite directions. There would be access driveways off each road onto the plot.

The condominium that Nails Harrelson wants to build would have an underground parking garage, commercial offices on the first floor, and then condominium living units on the upper levels. The condominiums would be sold, not rented. Harrelson assumes that all units will be sold quickly.

By law, no building over four stories can be constructed. Harrelson is willing to construct a building that has first-floor commercial space and just three more stories for living units.

He understands that he may ultimately need to limit the condominium area to two stories, however. A story in the building would contain 62,500 square feet—250 feet on a side.

Before construction can begin, the bar would have to be torn down and a large hole excavated for the parking garage. The foundation walls would then be put up. Then the access driveways and paving would be laid.

Condominium Plans

Nails Harrison plans to construct a spacious and appealing building. He will offer two condominium types: Upscale and Traditional.

- Assuming that three floors of condominiums are built, Harrelson would like to have between 70–75 condo units.
- For balance, Harrelson feels that at least 30 Upscale units and at least 30 Traditional units should be offered. The number of Upscale units must be at least as many as the number of Traditional units.
- Upscale condo units would sell for $300,000, contain 3,000 square feet of space, and have many amenities.
- Traditional condo units would sell for $150,000 and contain 2,000 square feet of space. They will have fewer amenities than the Upscale condos.
- The purchaser of an Upscale condo will be allotted three parking spaces in the underground garage, but the purchaser of a Traditional condo will get only two spaces, space permitting.

Retail/Office Space Plans

First-floor space could be used for retail businesses or for office space. Harrrelson is open to offers.

- First-floor units would be sold as roughed-in construction—four walls with electrical and water hookups. A buyer would have to finish the unit to suit their commercial purpose.
- Some of the first-floor space would have to be set aside for the building's heating and AC units and for other non-saleable spaces.
- Harrelson plans to create as many as 12 commercial units.
- Each commercial unit would contain 5,000 square feet and sell for $100,000, but a buyer could purchase several units and combine them. Harrelson wants a minimum of five commercial units, however.
- Each commercial unit would get five parking spaces in the underground garage. (Main floor units and condos will compete for parking spaces, a shared resource. Thus, the number of available parking spaces will impact your allocation decision for condo unit type.)

Building Construction Costs

Nails Harrelson has broken down construction costs as follows:

- Site preparation—demolition and excavation for the parking garage—would cost $1,000,000.
- The cost of grading, paving, and painting lines in the parking garage would be $10 a square foot. Harrelson plans to put in as many as 275 parking spaces, and each parking space would take up 200 square feet. There will also need to be open space so cars can drive in and out.

- The superstructure cost—the cost to put up the building's steel skeleton, outside walls, stairways and elevator, and so forth—will be $1.2 million per story.
- Finishing the condo units on each floor will vary. Harrelson thinks that an Upscale condo will cost $45 a square foot, and a Traditional condo will cost $40 a square foot.
- Construction costs for first-floor unit commercial space will be $10 a square foot.
- The roof, which will be installed by a special subcontractor, will cost $200,000.

Financing

Typically, real estate developers borrow most of the cost of construction and then pay back the loan as they sell units. That is Nails Harrelson's plan. He has a deal with the local bank in which the bank will lend 95% of the total construction funds at 10% interest. Harrelson will need to finance the remaining 5% of the project with his own money.

For all construction projects, Nails Harrelson has this profit goal: Net income after taxes should be as high as possible, of course, but at least 5% of total revenue from selling units. Nails has done projects that are not that profitable, of course. What real estate developer has not, when faced with hard times? But 5% seems like a reasonable minimum profit goal to him.

Harrelson wants an answer to this question: *How many of each type of condo and how many commercial units should be built?* You must model two situations: a Base Case (the parameters of which have just been described) and an alternative Extension Case for a downsized version (the parameters of which will be given shortly). You will use the Solver as your decision-support modeling tool. After you have finished the Base Case, you will modify its spreadsheet to create the Extension Case, which will let you play "what if" with the decision variables.

Nails Harrelson will use your results when he negotiates building the project with the Town Council. Ideally, he'd like to build according to his own goals, but he knows he may have to be flexible. However, he does not want to agree to build an unprofitable downsized building. Your case models will give Harrelson the information he needs to negotiate intelligently. In addition, you will write a memorandum that describes the problem and recommends a course of action. Your memorandum will include (as attachments) printouts of the Solver spreadsheets and reports documenting each case model. You will give your instructor your memo, printouts, and your disk or CD.

Assignment 1 Creating a Spreadsheet for Decision Support

In this assignment, you will produce a spreadsheet that models the business decision. In Assignment 1A, you will make a Solver spreadsheet to model the Base Case. In Assignment 1B, you will make a Solver spreadsheet to model the Extension Case. In Assignment 2, you will use the spreadsheet models to develop information needed to recommend the best unit mix, and then you will document your recommendation in a memorandum to Nails Harrelson. In Assignment 3, you will give your recommendation in an oral presentation.

Next, you will create the spreadsheet models of the construction design decision. Your spreadsheet should have the following sections:

- CHANGING CELLS
- CONSTANTS

- CALCULATIONS
- INCOME STATEMENT

Your spreadsheets will also include the decision constraints.

A Base Case spreadsheet skeleton is available to you, so you need not type in the skeleton. *To access the spreadsheet skeleton, go to your Data files, select Case 9, and then select **SPRUCE1.xls**.*

Assignment 1A: Creating the Spreadsheet—Base Case

You will model the construction problem described. Your model, when run, will tell Harrelson how many of each kind of unit to include in the building.

A discussion of each spreadsheet section follows. The discussion is about: (1) how each section should be set up, and (2) the logic of the sections' cell formulas.

CHANGING CELLS Section

Your spreadsheet should have the changing cells shown in Figure 9-1.

	A	B	C
1	BLUE SPRUCE DEVELOPMENT PLAN -- BASE CASE		
2			
3	CHANGING CELLS		
4	NUMBER OF UPSCALE CONDOS		1
5	NUMBER OF TRADITIONAL CONDOS		1
6	NUMBER OF RETAIL UNITS		1

Figure 9-1 CHANGING CELLS section

You are asking the Solver to compute the number of each type of condominium to include and the number of retail units to include in the first floor. Start with a 1 in each cell. The Solver will change each 1 as it computes the answer. Do not allow the Solver to recommend a fraction of a unit.

CONSTANTS Section

Your spreadsheet should have the constants shown in Figure 9-2. An explanation of the line items follows the figure.

	A	B	C
8	**CONSTANTS**		
9	TAX RATE		0.3
10	SELLING PRICE:		--
11	UPSCALE CONDO		300000
12	TRADITIONAL CONDO		150000
13	RETAIL UNIT		100000
14	SQUARE FOOTAGE:		--
15	UPSCALE CONDO		3000
16	TRADITIONAL CONDO		2000
17	RETAIL UNIT		5000
18	UNDERGROUND PARKING SPACE		200
19	CONSTRUCTION COST PER SQUARE FOOT:		--
20	UPSCALE CONDO		45
21	TRADITIONAL CONDO		40
22	RETAIL UNIT		10
23	UNDERGROUND PARKING SPACE		10
24	INFRASTRUCTURE COST PER STORY		1200000
25	SQUARE FOOTAGE IN A STORY		62500
26	MAXIMUM NUMBER OF:		--
27	PARKING PLACES		275
28	STORIES FOR CONDOS		3
29	RETAIL UNITS		12
30	CONDOS		75
31	PARKING SPOTS PER:		--
32	UPSCALE CONDO		3
33	TRADITIONAL CONDO		2
34	RETAIL UNIT		5
35	MAXIMUM SQUARE FOOTAGE FOR CONDOS		187500
36	CONSTRUCTION LOAN INTEREST RATE		0.1
37	GROUND PREPARATION COST		1000000
38	COST OF ROOF		200000

Figure 9-2 CONSTANTS section

- TAX RATE: Positive net income before taxes is taxed at 30%. There is no tax on losses.
- SELLING PRICE: All units will be sold and none will be rented. The different selling prices are shown. Assume that all units that are built will be sold. Parking spaces are included in the purchase price.
- SQUARE FOOTAGE: Each kind of unit has a different square footage. A parking space will take up 200 square feet in the underground garage.
- CONSTRUCTION COST PER SQUARE FOOT: Each kind of unit has a different average cost per square foot, as shown.
- INFRASTRUCTURE COST PER STORY: It will cost $1.2 million to erect each building story.
- SQUARE FOOTAGE IN A STORY: A condo story will have 62,500 square feet of usable space.

- MAXIMUM NUMBER OF: There can be no more than 275 parking spaces. There can be no more than three stories for condos. There can be no more than 12 retail units in the first floor. There can be no more than 75 condos built.
- PARKING SPACES PER: The purchaser of an Upscale condo gets three parking spaces; the purchaser of a Traditional condo gets two parking spaces. Each 5,000-square-foot-retail unit gets five underground parking spaces.
- MAXIMUM SQUARE FOOTAGE FOR CONDOS: There can be no more than three floors for condos, so no more than 187,500 square feet can be used for condos.
- CONSTRUCTION LOAN INTEREST RATE: The bank will lend 95% of the construction cost at 10% interest.
- GROUND PREPARATION COST: It will cost $1 million for demolition, debris removal, and garage excavation.
- COST OF ROOF: The cost of the roof will be $200,000.

CALCULATIONS Section

Your spreadsheet should calculate the amounts for the line items shown in Figure 9-3. The amounts will be used in the INCOME STATEMENT section or in the CONSTRAINTS. Calculations are based on values in the CHANGING CELLS section and/or in the CONSTANTS section and/or on other calculated values. An explanation of the line items follows the figure.

	A	B	C
40	**CALCULATIONS**		
41	SQUARE FOOTAGE USED:		--
42	UPSCALE CONDO		
43	TRADITIONAL CONDO		
44	RETAIL UNIT		
45	TOTAL SQUARE FOOTAGE USED FOR CONDOS		
46	NUMBER OF STORIES USED FOR CONDOS		
47	INFRASTRUCTURE COST FOR STORIES		
48	UNDERGROUND PARKING SPACES NEEDED:		--
49	CONDOS		
50	RETAIL UNITS		
51	REVENUE:		--
52	UPSCALE CONDO		
53	TRADITIONAL CONDO		
54	RETAIL UNIT		
55	CONSTRUCTION COSTS:		--
56	UPSCALE CONDO		
57	TRADITIONAL CONDO		
58	RETAIL UNIT		
59	UNDERGROUND PARKING SPACE		
60	NET INCOME TO REVENUE RATIO		
61	NUMBER OF CONDOS BUILT		
62	TOTAL UNDERGROUND PARKING SPACES		

Figure 9-3 CALCULATIONS section

- SQUARE FOOTAGE USED: Square footage used for each kind of unit depends on the number of each kind of unit built and the square footage per unit.
- TOTAL SQUARE FOOTAGE USED FOR CONDOS: This is the sum of square footage used for Upscale and Traditional condos.
- NUMBER OF STORIES USED FOR CONDOS: If square footage used is less than or equal to 62,500, then only one story will be needed. If square footage is between 62,501 and 125,000 square feet, two stories will be needed. If square footage is between 125,001 and 187,500 square feet, then three stories will be needed. If square footage is above 187,500, four stories would be needed. (*Hint*: Your calculation should allow for four stories to give the Solver room to work as it tries out possible answers.)
- INFRASTRUCTURE COST FOR STORIES: This is a function of the number of stories and the cost per story.
- UNDERGROUND PARKING SPACES NEEDED: This is a function of the number of each kind of unit and the parking spaces for that kind of unit.
- REVENUE: Assume that all units built will be sold. Revenue values are a function of the number of each kind of unit and the selling price of each.
- CONSTRUCTION COSTS: Construction cost values are a function of the number of each kind of unit, the square footage for each, and the cost per square foot for each.
- NET INCOME TO REVENUE RATIO: This is the ratio of net income after taxes to total revenue.
- NUMBER OF CONDOS BUILT: This is the total of Upscale and Traditional units built.
- TOTAL UNDERGROUND PARKING SPACES: This is a function of the number of each kind of unit built and the number of spaces for each kind of unit.

INCOME STATEMENT Section

Harrelson thinks that all units can be built and sold in one year, so you make an income statement for the project for one year. Compute the project's net income after taxes, as shown in Figure 9-4. Line items are discussed after the figure.

	A	B	C
64	**INCOME STATEMENT**		
65	TOTAL REVENUE		
66	COSTS:		--
67	CONSTRUCTION COST		
68	INFRASTRUCTURE COST		
69	GROUND PREPARATION + ROOF COSTS		
70	TOTAL COSTS		
71	NET INCOME BEFORE INTEREST AND TAXES		
72	INTEREST EXPENSE		
73	NET INCOME BEFORE TAXES		
74	INCOME TAX EXPENSE		
75	NET INCOME AFTER TAXES		

Figure 9-4 INCOME STATEMENT section

- TOTAL REVENUE: This is the total from selling all the condos and commercial units built.
- CONSTRUCTION COST: This is the total cost of building condos and commercial units and finishing the related parking spaces.
- INFRASTRUCTURE COST: This is the infrastructure cost for the condo stories and the ground floor story.
- GROUND PREPARATION + ROOF COSTS: This is the sum of the ground preparation and roof costs.
- TOTAL COSTS: This is the sum of construction, infrastructure, ground preparation, and roof costs.
- NET INCOME BEFORE INTEREST AND TAXES: This is the difference between total revenue and total costs.
- INTEREST EXPENSE: The bank will lend 95% of total costs, computed in this income statement. Interest expense is 10%, a constant.
- NET INCOME BEFORE TAXES: This is net income before interest and taxes, less interest expense.
- INCOME TAX EXPENSE: Income taxes are paid at the tax rate on positive net income before taxes. If net income before taxes is zero or negative, income tax expense is zero.
- NET INCOME AFTER TAXES: This is net income before taxes, less income tax expense.

Constraints and Running the Solver

Determine the constraints. Enter the Base Case decision constraints, using the Solver. Run the Solver. Make an Answer Report when the Solver says that a solution has been found that satisfies the constraints.

When you are finished, print the entire workbook, including the Answer Report. Save the Base Case Solver spreadsheet file (File—Save; **SPRUCE1.xls** should be the filename). Then, to prepare for the Extension Case, use File—Save As to make a new spreadsheet. (**SPRUCE2.xls** should be the filename.)

Assignment 1B: Creating the Spreadsheet—Extension Case

From reliable sources, Nails Harrelson has some idea of how the influential Town Council members will want his project downsized. He might have to adhere to these rules:

- No more than two stories of condos—hence, no more than 125,000 square feet for condos.
- To curb downtown traffic congestion, no more than 50 condominium units can be built.
- Each condominium floor will have to be made more accessible to handicapped people—facilities will have to be better than those actually required by law. This will increase the infrastructure cost of a condominium story to $1.3 million.

Nails Harrelson would react to these requirements as follows:

- He would build at least 20 Upscale and 20 Traditional condos, and at least as many Upscale as Traditional. He would want to build at least 45 condos in total, up to the 50 allowed.
- He would increase the number of Upscale parking spaces to four. This measure makes the Upscale condo more desirable, so the price of a unit would rise to $310,000.

- He would increase the number of Traditional parking spaces to three. The selling price of a unit would rise by $10,000.
- The Town Council apparently will not try to control the number of retail units. So, Nails would try to build and sell up to 15 commercial units, at 4,000 square feet apiece, and sell at a price of $95,000 each.
- Harrelson would drop his 5% profit goal. Your model should compute the ratio of net income to revenue for Harrelson's information, but no profit goal should be assumed in your model as a constraint.

Nails Harrelson wants to know how much money—if any—will be made under the rumored restrictions. He does not want to build a money-losing project.

Modify the Extension Case spreadsheet (**SPRUCE2.xls**) and related constraints to reflect the expected Town Council modifications. Run the Solver. Ask for an Answer Report when the Solver says that a solution has been found that satisfies the constraints. When you are done, print the entire workbook, including the Solver Answer Report. Save the worksheet when you are finished. Close the file and exit Excel.

Assignment 2 Using the Spreadsheet for Decision Support

You have built the Base Case and the Extension Case models because you want to know the profitability of each configuration. You will now complete your work by: (1) using the worksheets and Answer Reports to gather the data needed to help Harrelson negotiate with Town Council, and (2) documenting your recommendation in a memorandum and (if your instructor specifies) an oral presentation.

Assignment 2A: Using the Spreadsheet to Gather Data

You have printed the spreadsheet and Answer Report sheet for each case. You can see the results for each configuration. You should summarize the key data in a table that will be included in your memorandum. The form of that table is shown in Figure 9-5.

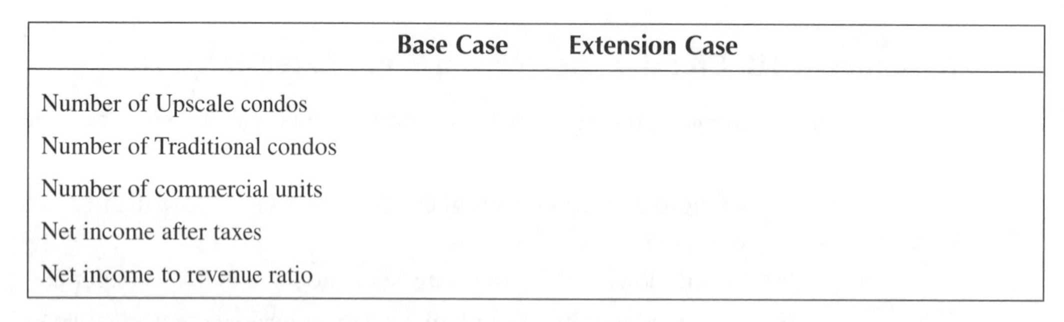

	Base Case	Extension Case
Number of Upscale condos		
Number of Traditional condos		
Number of commercial units		
Net income after taxes		
Net income to revenue ratio		

Figure 9-5 Format of table to insert in memorandum

Assignment 2B: Documenting Your Recommendation in a Memorandum

Use MS Word to write a brief memorandum to Harrelson about the results of your analysis. Observe the following requirements:

1. Your memorandum should have a proper heading (DATE / TO / FROM / SUBJECT). You might want to use a Word memo template (**File**, click **New**, click **On my computer** in the Templates section, click the **Memos** tab, choose **Contemporary Memo**, and then click **OK**).
2. Briefly state the problem and the decision to be made, but do not provide background—Nails Harrison knows the background. You should briefly state your analytical method and state the results of the analysis. Give him your negotiation advice, which should be keyed to the net income after taxes results.
3. Support the recommendation graphically by including a summary table in your Word memo, as shown in Figure 9-5. Make the table by following the procedure described next.

Enter a table into the Word memorandum, using the following procedure:

1. Select the **Table** menu option, click **Insert**, and then click **Table**.
2. Enter the number of rows and columns.
3. Select **AutoFormat** and choose **TableGrid1**.
4. Select **OK**, and then select **OK** again.
5. Once the table is in place, merely enter data directly into its cells.

Case 9

ASSIGNMENT 3 GIVING AN ORAL PRESENTATION

Your instructor might request that you also present your analysis and recommendations in an oral presentation. If so, assume that Nails Harrelson is very impressed by your analysis, and that he asks you to give a presentation to his management team and to the Town Council. Prepare to explain your analysis and recommendation to the group in 10 minutes or fewer. Use visual aids or handouts that you think are appropriate. Tutorial E has guidance on how to prepare and give an oral presentation.

DELIVERABLES

Assemble the following deliverables for your instructor:

1. Printout of your memorandum
2. Spreadsheet printouts
3. Disk or CD, which should have your Word memorandum and Excel spreadsheet files

Staple the printouts together, with the memorandum on top. Hand-write your instructor a note, stating the names of your Base Case and Extension Case *.xls* files.

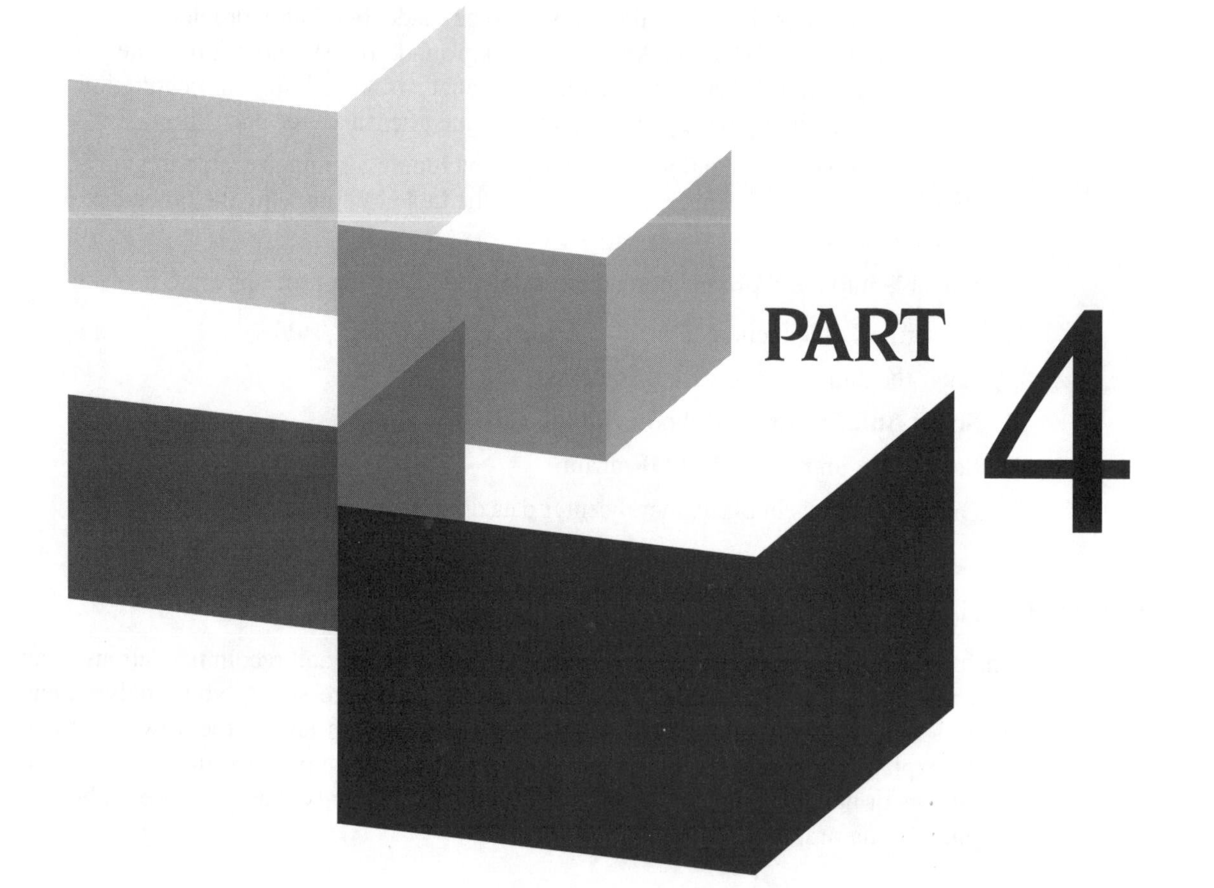

PART 4

Decision Support Case Using Basic Excel Functionality

Wonderful Software's Pension Plan Decision

10
CASE

Decision Support Using Excel

Preview

Wonderful Software Company is considering making changes to its expensive retirement plan. In this case, you will make a spreadsheet that will let management decide whether the proposed changes are to the company's advantage.

Preparation

- Review spreadsheet concepts discussed in class and/or in your textbook.
- Complete any exercises that your instructor assigns.
- Complete any parts of Tutorials C and D that your instructor assigns, or refer to them as necessary.
- Review file-saving procedures for Windows programs. These are discussed in Tutorial C.
- Refer to Tutorial E as necessary.

Background

Wonderful Software Company has been very successful since the early days of personal computing—more than 20 years ago! Because the company has been in business for so many years, it has a sizable number of retirees; currently, the company has 50,000 employees and has 10,000 retirees. Looking ahead to baby-boomer retirements, the company's management knows that the number of retirees will increase greatly and pension expenditures will be very heavy. Presently, the company pays retirees in two ways:

1. Wonderful Software runs a traditional, privately funded retirement program. Each retiree is paid a yearly pension, and every year the company puts an amount into the pension plan sufficient to pay retirees for that year. That amount equals the current year pension payment times the number of retirees. The number of retirees equals the number of retirees at the start of the year, plus those who retire in the year, less those who die in the year.
2. In addition to the pension plan, the company also contributes to employees' 401K plans. All employees participate in this plan, which is administered by a famous mutual fund company. Under this plan, each employee contributes 6% and the company contributes 4% of the employee's pay into the plan. In effect, this 4% is "free" money for the employee. Each employee's 401K investment mounts up during his or her employment years and is dispersed to the employee upon retirement.

Management thinks that the traditional pension plan is too expensive, and they want to discontinue the plan for current employees. Under the new plan, current employees would get just the 401K plan when they retire. However, those who have already retired would continue to get both the traditional pension and the 401K funds.

Management knows that employees who have worked for several years feel that they have "earned" the right to receive a pension when they retire. In this company's retirement plan, no legal obligation actually exists, but that will likely not change employees' expectations. Dropping the pension plan could create a significant employee-relations problem. Management has proposed two inducements to make the change acceptable to employees:

1. The company could increase its contribution to the 401K plan. Under tax law, the company is allowed to contribute funds up to 6% of employee pay. Management would raise the 401K contribution rate from 4%, possibly all the way to 6%.
2. The company could put a lump sum amount into the 401K mutual fund plan for each current employee, to make up for those pension payments that employees think they have earned. At least $500 million, possibly more, will be deposited with the mutual fund for the benefit of existing employees' 401K plans.

Management will make the change if it is financially advantageous. That is where you come in. You are asked to make a spreadsheet that compares 10 years' pension and 401K expenses under the current and the proposed plans. If the proposed plan is less expensive over 10 years (2007-2016), the company will adopt it; otherwise, the company will maintain the existing retirement payment plan.

Assignment 1 Creating a Spreadsheet for Decision Support

In this assignment, you will produce a spreadsheet that models the business problem, using the scenarios in Figure 10-10. Then, in Assignment 2, you will write a memorandum to management that explains your findings and recommends a course of action. In addition, in Assignment 3, you might be asked to prepare an oral presentation of your analysis and recommendation.

Now, you will create the retirement expense forecast in Excel. Your spreadsheet should have three worksheets. You will be given some hints on how the sections in each worksheet should be set up before entering cell formulas. Your spreadsheet should have these worksheets:

- **TOP**—to summarize results and compare alternatives
- **FUND**—to compute pension fund payments under each alternative
- **401K**—to compute 401K payments under each alternative

Each of the worksheets is discussed next. *The spreadsheet skeleton is available to you, so you need not type in the skeleton if you do not want to do so.* To access the spreadsheet skeleton, go to your Data files. Select Case 10 and then select **401K.xls**.

TOP Worksheet

Your worksheet should have these sections:

- CONSTANTS
- INPUTS
- CALCULATIONS

A discussion of each section follows.

CONSTANTS Section

Your spreadsheet should have the constants shown in Figure 10-1. An explanation of the line items follows the figure.

	A	B	C	D
1	WONDERFUL SOFTWARE'S PENSION CONVERSION			
2				
3	CONSTANTS	2006	2007	2008
4	CURRENT 401K MATCHING %	0.04	NA	NA
5	NUMBER OF RETIREES	10000	NA	NA

Figure 10-1 CONSTANTS section

- CURRENT 401K MATCHING %: The current matching percent is 4% of employee pay.
- NUMBER OF RETIREES: Currently, the company has 10,000 retired people receiving pension payments.

INPUTS Section

Your worksheet should have the inputs shown in Figure 10-2. Only the first few of the 10 analysis years are shown. Your instructor might tell you to apply Conditional Formatting to the input cells, so that out-of-bounds values are highlighted in some way (for example, the entry is shown in red and/or in boldface type). If so, your instructor might provide a tutorial on Conditional Formatting. Or, your instructor might ask you to refer to Excel Help. An explanation of the line items follows the figure.

	A	B	C	D
7	INPUTS	2006	2007	2008
8	NEW 401K MATCHING %		NA	NA
9	% INCREASE IN EMPLOYEES		NA	NA
10	% INCREASE IN AVG SALARY		NA	NA
11	RETIREE DEATH RATE	NA		
12	401K MAKEUP		NA	NA

Figure 10-2 INPUTS section

- NEW 401K MATCHING %: The user would enter an amount up to 6% (.06). This amount would apply to all 10 years.
- % INCREASE IN EMPLOYEES: The user would enter a yearly increase in employees. For example, if the number of employees was expected to increase 1% each year, .01 would be entered. If the number of employees was expected to decrease 2%, -.02 would be entered. This amount would apply to all 10 years.
- % INCREASE IN AVG SALARY: The user would enter a yearly increase in salary. For example, if the average salary was expected to increase 1% each year, .01 would be entered. If the average salary was expected to decrease 2%, -.02 would be entered. This amount would apply to all 10 years.
- RETIREE DEATH RATE: If 8% of retirees are expected to die in a year, .08 would be entered. A value would be entered for each of 10 years.
- 401K MAKEUP: This is the one-time lump-sum amount that the company would deposit with the mutual fund. At least $500 million is scheduled, but any amount over that could be entered.

CALCULATIONS Section

This section is the spreadsheet body. Calculate the values shown in Figure 10-3. An explanation of the line items follows the figure.

	A	B	C	D
14	CALCULATIONS	2006	2007	2008
15	CURRENT PLAN COMPENSATION:	--	--	--
16	PENSION	NA		
17	401K	NA		
18	TOTAL	NA		
19				
20	NEW PLAN COMPENSATION:	--	--	--
21	PENSION	NA		
22	401K	NA		
23	TOTAL	NA		
24				
25	INVESTMENT / SAVINGS			
26				
27	CUMULATIVE SAVINGS			

Figure 10-3 CALCULATIONS section

- CURRENT PLAN COMPENSATION: Pension payments in each year are echoed to here from the **FUND** worksheet. The 401K payments in each year are echoed to here from the **401K** worksheet. The total in each year is the sum of pension and 401K payments.
- NEW PLAN COMPENSATION: Pension payments in each year are echoed to here from the **FUND** worksheet. The 401K payments in each year are echoed to here from the **401K** worksheet. The total in each year is the sum of pension and 401K payments.

- INVESTMENT / SAVINGS: In 2006, the investment amount is the value of the lump-sum makeup, which can be echoed to here. The 2006 amount would show as a negative amount. In each of the next 10 years, the savings amount is current plan total payments, less new plan total payments.
- CUMULATIVE SAVINGS: Cumulative savings in a year is the accumulated net savings. In 2006, this number is negative—there has only been an investment for the catch-up amount. In 2007, the amount will be the investment, offset by any 2007 savings. In 2008, the amount will be the 2007 cumulative amount offset by any 2008 savings, and so on, for the rest of the 10 years.

Values can be echoed from one worksheet to another using exclamation point notation. For example, suppose that you have a value in cell B4 of a worksheet named BottomSheet. You want that value to appear in cell A6 of a worksheet called Sheet1. In cell A6, you would put the formula *=BottomSheet!B4*.

The astute student will note the importance of the cumulative savings amount. If the number is still negative at the end of the 10-year period, then savings do not justify the lump sum investment, and adopting the new plan is not feasible, at least not with the inputs used.

Case 10

FUND Worksheet

Your worksheet should have these sections:

- CONSTANTS
- INPUTS
- CALCULATIONS

A discussion of each section follows.

CONSTANTS Section

Your spreadsheet should have the constants shown in Figure 10-4. Only the first two of the 10 analysis years are shown. An explanation of the line items follows the figure.

	A	B	C	D
1	WONDERFUL SOFTWARE'S PENSION CONVERSION			
2				
3	CONSTANTS	2006	2007	2008
4	CURRENT 401K MATCHING %	0.04	NA	NA
5	NEW RETIREES EACH YEAR	NA	1500	1500
6	EXPECTED PAYMENT TO RETIREE	NA	40000	41000

Figure 10-4 CONSTANTS section

- CURRENT 401K MATCHING %: This constant is echoed from the **TOP** worksheet to this cell. The value applies to all years.
- NEW RETIREES EACH YEAR: Management has estimated the number of new retirees in each year. The values for the first 2 of the 10 analysis years are shown. Management estimates 1,500 retirees per year in the first years, then 2,000 retirees per year for the next 8 years.
- EXPECTED PAYMENT TO RETIREE: Management expects to pay a retiree $40,000 in 2007, with the amount increasing $1,000 each year for 9 years.

INPUTS Section

Your **FUND** worksheet should have the inputs shown in Figure 10-5. Only the first few of the 10 analysis years are shown. Your instructor might tell you to apply Conditional Formatting to the input cells, so that out-of-bounds values are highlighted. An explanation of the line items follows the figure.

	A	B	C	D
8	INPUTS	2006	2007	2008
9	NEW 401K MATCHING %		NA	NA
10	DEATH RATE	NA		

Figure 10-5 INPUTS section

- NEW 401K MATCHING %: This input should be echoed from the **TOP** worksheet and can be referenced in cell formulas in this worksheet. The amount applies to all years.
- DEATH RATE: This input should be echoed from the **TOP** worksheet and can be referenced in cell formulas in this worksheet.

Warning! If you want to change an input value, change it in the **TOP** worksheet and echo the value to this worksheet!

CALCULATIONS Section

Calculate the values shown in Figure 10-6. An explanation of the line items follows the figure.

	A	B	C	D
12	CALCULATIONS	2006	2007	2008
13	NUMBER OF RETIREES -- EXISTING PLAN:	--	--	--
14	ADD: NEW RETIREES IN PLAN	NA		
15	LESS: DEATHS OF RETIREES	NA		
16	NUMBER OF RETIREES IN PLAN	10000		
17				
18	NUMBER OF RETIREES -- NEW PLAN:	--	--	--
19	ADD: NEW RETIREES IN PLAN	NA		
20	LESS: DEATHS OF RETIREES	NA		
21	NUMBER OF RETIREES IN PLAN	10000		
22				
23	PENSION EXPENSE:	--	--	--
24	EXISTING PLAN	NA		
25	NEW PLAN	NA		

Figure 10-6 CALCULATIONS section

- NUMBER OF RETIREES—EXISTING PLAN: ADD NEW RETIREES IN PLAN: This is a constant that can be echoed here.
- NUMBER OF RETIREES—EXISTING PLAN: LESS DEATHS OF RETIREES: This is a function of the death rate (an input) and the number of retirees in the plan at the end of the prior year. Assume that retirees do not die in the year in which they retire.
- NUMBER OF RETIREES—EXISTING PLAN: NUMBER OF RETIREES IN PLAN: The 2006 value is shown. For a year, this equals the number of the retirees in the plan at the end of the prior year, plus new retirees, less deaths of retirees.

- NUMBER OF RETIREES—NEW PLAN: ADD NEW RETIREES IN PLAN: There will be no new retirees. You can hard-code 10 zeroes here.
- NUMBER OF RETIREES—NEW PLAN: LESS DEATHS OF RETIREES: This is a function of the death rate (an input) and the number of retirees in the plan at the end of the prior year.
- NUMBER OF RETIREES—NEW PLAN: NUMBER OF RETIREES IN PLAN: The 2006 value is shown. For a year, this equals the number in the plan at the end of the prior year, plus new retirees, less deaths of retirees.
- PENSION EXPENSE: EXISTING PLAN: This is a function of the expected payment in the year to a retiree (a constant) and the number of retirees (a calculation). This amount is echoed to the **TOP** worksheet.
- PENSION EXPENSE: NEW PLAN: This is a function of the expected payment in the year to a retiree (a constant) and the number of retirees (a calculation). This amount is echoed to the **TOP** worksheet.

Case 10

401K Worksheet

Your worksheet should have these sections:

- CONSTANTS
- INPUTS
- CALCULATIONS

A discussion of each section follows.

CONSTANTS Section

Your **401K** worksheet should have the constant shown in Figure 10-7. Only the first few of the 10 analysis years are shown. An explanation of the line item follows the figure.

	A	B	C	D
1	WONDERFUL SOFTWARE'S PENSION CONVERSION			
2				
3	CONSTANTS	2006	2007	2008
4	CURRENT 401K MATCHING %	0.04	NA	NA

Figure 10-7 CONSTANTS section

- CURRENT 401K MATCHING %: This constant is echoed from the **TOP** worksheet to this cell. The value applies to all years.

INPUTS Section

Your worksheet should have the inputs shown in Figure 10-8. Only the first two of the 10 analysis years are shown. Your instructor might tell you to apply Conditional Formatting to the input cells, so that out-of-bounds values are highlighted. An explanation of the line items follows the figure.

	A	B	C	D
6	INPUTS	2006	2007	2008
7	NEW 401K MATCHING %		NA	NA
8	% INCREASE IN EMPLOYEES		NA	NA
9	% INCREASE IN AVG SALARY		NA	NA

Figure 10-8 INPUTS section

- NEW 401K MATCHING %: This input should be echoed from the **TOP** worksheet and can be referenced in cell formulas in this worksheet. The amount applies to all years.
- % INCREASE IN EMPLOYEES: This input should be echoed from the **TOP** worksheet and can be referenced in cell formulas in this worksheet. The amount applies to all years.
- % INCREASE IN AVG SALARY: This input should be echoed from the **TOP** worksheet and can be referenced in cell formulas in this worksheet. The amount applies to all years.

Warning! If you want to change an input value, change it on the **TOP** worksheet and echo the value to this worksheet!

CALCULATIONS Section

Calculate the values shown in Figure 10-9. An explanation of the line items follows the figure.

	A	B	C	D
11	CALCULATIONS	2006	2007	2008
12	NUMBER OF EMPLOYEES	50000		
13	AVERAGE SALARY	60000		
14	SALARY EXPENSE	NA		
15	CURRENT 401K EXPENSE	NA		
16	NEW PLAN 401K EXPENSE	NA		

Figure 10-9 CALCULATIONS section

- NUMBER OF EMPLOYEES: This amount is a function of the number of employees in the prior year and the percentage increase expected, a constant. For example, if there are 50,000 employees to start the year and the percentage of increase is 5%, then the number of employees in the year would be 50,000 plus 5% of 50,000, or 52,500. In the next year, the number of employees would be 52,500 plus 5% of 52,500, and so on, for succeeding years.
- AVERAGE SALARY: The average salary was $60,000 in 2006. This amount in a later year is a function of the average salary in the prior year and the percentage increase expected, a constant. For example, if the average salary is $60,000 to start the year and the percentage increase is 10%, then the average salary in the year would be $60,000 plus 10% of $60,000, or $66,000. In the next year, the average salary would be $66,000 plus 10% of $66,000, and so on, for succeeding years.
- SALARY EXPENSE: This amount is a function of the number of employees in the year and the average salary in the year.
- CURRENT 401K EXPENSE: This amount is a function of the salary expense and the current 401K matching rate, a constant. This value is echoed to the **TOP** worksheet.
- NEW PLAN 401K EXPENSE: This amount is a function of the salary expense and the new plan's 401K matching rate, an input. This value is echoed to the **TOP** worksheet.

ASSIGNMENT 2 USING THE SPREADSHEET FOR DECISION SUPPORT

You will now complete the case by: (1) using the spreadsheet to gather the data needed to decide whether the pension plan change makes financial sense, (2) documenting your recommendation in a memorandum, and (3) if your instructor specifies, giving an oral presentation.

Assignment 2A: Using the Spreadsheet to Gather Data

You have built the spreadsheet to compare retirement payments under the current pan and a revised plan. Management thinks that the revised plan will be better financially, but wants to see numbers.

You should run four "what-if" scenarios, summarized in Figure 10-10. A discussion of each scenario follows the figure.

	Cheap	Realistic	Pessimistic	Optimistic
New 401K				
matching %	.05	.06	.06	.06
% increase in employees	.00	.01	.02	–.01
% increase in average salary	.00	.02	.03	–.01
Retiree death rate	.08	.08	.08	.08
401K makeup	$500,000,000	$750,000,000	$1,000,000,000	$500,000,000

Figure 10-10 Four scenario inputs

1. ***Cheap scenario***: Management thinks they might be able to gain employee goodwill if they match only 5% of pay, keep the workforce stable, offer no pay raises, and create a $500 million 401K makeup. The retiree death rate would be unchanged in all scenarios.
2. ***Realistic scenario***: Management assumes that they will need to match 6% of employee pay, increase the workforce and salaries somewhat, and create a $750 million 401K makeup.
3. ***Pessimistic scenario***: In this scenario, management would match 6% of employee pay, increase the payroll and average salary even more, and contribute $1 billion to the 401K makeup.
4. ***Optimistic scenario***: In this scenario, management would match 6% of employee pay. However, productivity increases actually allow a reduction in the workforce and average salary, and only $500 million is needed for the 401K makeup amount.

In any case, management wants to know what the cumulative savings will be at the end of 10 years. In the early years, the cumulative savings will be negative because of the large 401K makeup amount. However, management hopes that the cumulative number will turn positive in some year. Management wants to know the year in which the savings do turn positive. That data can be summarized in a table similar to the one shown in Figure 10-11.

	Cheap	Realistic	Pessimistic	Optimistic
Cumulative 10-year savings				
Year in which savings turn positive (if any)				

Figure 10-11 Scenario results

You should manually enter the input values in the **TOP** worksheet—these should echo to the other worksheets by cell formula. Note the target outputs in each scenario using a separate sheet of paper.

When you are done gathering data, print the entire workbook (with any input value combination entered). Then, save the spreadsheet (File—Save).

Assignment 2B: Documenting Your Recommendation in a Memorandum

Now you will document your recommendation in a memorandum. Management wants the answers to these questions:

1. Is there any situation in which the current plan would be preferable? Or is it a no-brainer: Go for the revised plan, no matter what?
2. Considering only those scenarios in which the revised plan is preferred, are these scenarios essentially the same in their financial impact? "Essentially the same" would mean less than $200 million in cumulative savings after 10 years. That is, how different are the scenarios from each other?
3. Going for the *Cheap* scenario might mean incurring some employee ill-will. Comparing the *Realistic* scenario to the *Cheap* scenario: Would it be cost-effective to adopt the *Cheap* scenario? Assume that it might be if the difference in cumulative savings exceeds $500 million dollars.
4. Acquiescing to the *Pessimistic* scenario's demands might mean incurring some significant expenses. Compare the *Realistic* scenario to the *Pessimistic* scenario: Should management fight hard to avoid the *Pessimistic* scenario? Assume that it probably would be worth it if the difference in cumulative savings exceeds $250 million.
5. Going for the *Optimistic* scenario would probably involve some serious outsourcing and other belt-tightening measures. Compare the *Realistic* scenario to the *Optimistic* scenario: Should management go for workforce and salary reductions? Assume that they should strongly consider doing that if the cumulative savings difference exceeds $500 million.
6. What is your bottom line on this? If you were the CEO, which scenario would you advocate?

Here is guidance on preparing your memorandum in MS Word:

- Your memorandum should have a proper heading (DATE / TO / FROM / SUBJECT). You might want to use a Word memo template (**File**, click **New**, click **On my computer** in the Templates section, click the **Memos** tab, choose **Contemporary Memo**, then click **OK**).
- Briefly outline the business situation. However, you need not provide much background—you can assume the CEO is familiar with the problem.

- Answer management's questions in the body of the memorandum.
- Support your claims graphically, by including a table like the one shown in Figure 10-11.

Enter the table into Word, using the following procedure:

1. Select the **Table** menu option, point to **Insert**, then click **Table**.
2. Enter the number of rows and columns.
3. Select **AutoFormat** and choose **Table Grid 1**.
4. Select **OK** and then select **OK** again.

ASSIGNMENT 3 GIVING AN ORAL PRESENTATION

Your instructor might request that you also present your results in an oral presentation. If so, assume that the CEO is impressed by your analysis and has asked you to give a presentation explaining your results to the company's senior management. Prepare to explain your forecast and results to the group in 10 minutes or fewer. Use visual aids or handouts that you think are appropriate. Tutorial E has guidance on how to prepare and give an oral presentation.

DELIVERABLES

1. Printouts of your memorandum
2. Spreadsheet printouts
3. Disk or CD, which should have your Word memo file and your Excel spreadsheet file

Staple the printouts together, with the memorandum on top. If there is more than one *.xls* file on your disk, write your instructor a note, stating the name of your model's *.xls* file.

PART 5

Integration Case: Using Access and Excel

The Stock Portfolio Restructuring

DECISION SUPPORT USING ACCESS AND EXCEL

PREVIEW

A mutual fund portfolio manager wants to restructure her fund using "troubled stock" buy-and-sell rules. You have been called in to automate the restructuring, using an Excel spreadsheet and Access database.

PREPARATION

- Review spreadsheet and database concepts discussed in class and/or in your textbook.
- Complete any exercises that your instructor assigns.
- Obtain the database file **INVEST.mdb** from your Data files in the Case 11 folder.
- Review any part of Tutorials A, B, C, or D that your instructor specifies, or refer to them as necessary.
- Review file-saving procedures for Windows programs. These are discussed in Tutorial C.
- Refer to Tutorial E as necessary.

BACKGROUND

Your friend Holly Smith has just taken over as manager of a mutual fund. The fund has several stocks, some of which have declined in price. Holly thinks that the fund needs to be restructured, and she wants to weed out stocks that are ripe for sale. With the proceeds, she will buy other stocks that appear to be undervalued in the stock market.

Holly has a database with a table for stocks the fund owns and another table for stocks the fund might buy. She wants to use the database and a spreadsheet to automate the portfolio restructuring. Knowing that you are proficient in Access and Excel, she has asked you for help.

Holly wants to follow the "troubled stocks" strategy in her restructuring. The strategy has the following rules for selling and buying stocks:

- Holly would sell a stock if its price has risen 25% or more since it was purchased. She reasons that such a stock's price will not increase in value much more and might go down, so why not sell it now and take the profit?
- Holly would use stock sale proceeds to buy stocks that are now selling for 75% or less than their highest price in the past 52 weeks. She reasons that a stock that has fallen that far might be ripe for an increase in price.
- The strategy also has rules for industry concentrations. She would not want to put more than 20% of the portfolio in stocks of companies in any one industry. However, she does not want trivial representation either, so she would want stocks of any industry to represent at least 8% of the portfolio.

The database **INVEST.mdb** has a table of data about stocks currently owned in the portfolio. The database also has a table of data about stocks that could possibly be purchased. The database tables are discussed next.

The OWNED Table

The OWNED stocks table shows data about the portfolio's stocks. The design of the table is shown in Figure 11-1. A discussion of the table's fields follows the figure.

OWNED : Table

	Field Name	Data Type
🔑	STOCK NAME	Text
	INDUSTRY	Text
	SHARES	Number
	AVERAGE PRICE	Number
	MARKET PRICE	Number

Figure 11-1 Design of the OWNED table

- STOCK NAME: The field shows the company's name.
- INDUSTRY: The field shows the company's industry.
- SHARES: The field shows the number of shares of the stock owned in the portfolio.
- AVERAGE PRICE: The field shows the average purchase price of the shares owned in the portfolio.
- MARKET PRICE: The field shows the current share price in the stock market.

There are many stocks in the portfolio. The first few records are shown in Figure 11-2.

OWNED : Table

STOCK NAME	INDUSTRY	SHARES	AVERAGE PRICE	MARKET PRICE
ACE	CAPITAL	1000	82	61
BCD	FINANCE	5000	76	36
BCDE	HEALTHCARE	1000	89	73
BCDEF	TECHNOLOGY	9000	72	84
BCDEFG	MINING	1000	91	70

Figure 11-2 Records in the OWNED table

For interpretation of the data in Figure 11-2, look at the record for Stock BCD. It is in the Finance industry, and the fund currently owns 5,000 shares. The average purchase price is $76 per share, which means that the book value of the investment in BCD is 5,000 * $76 = $380,000. The current market price of BCD is only $36 per share, however, which means that the stocks would fetch only 5,000 * $36 = $180,000 if they were sold today.

The COULD BUY Table

The COULD BUY table shows data about stocks that Holly thinks are interesting and worth following. The design of the table is shown in Figure 11-3. A discussion of the table's fields follows the figure.

COULD BUY : Table

Field Name	Data Type
STOCK NAME	Text
INDUSTRY	Text
52 WEEK HIGH	Number
MARKET PRICE	Number

Figure 11-3 Design of the COULD BUY table

- STOCK NAME: The field shows the company's name.
- INDUSTRY: The field shows the company's industry.
- 52 WEEK HIGH: The field shows the stock's highest market price in the past 52 weeks.
- MARKET PRICE: The field shows the current share price in the stock market.

Holly follows several stocks. The records for the first few stocks are shown in Figure 11-4.

COULD BUY : Table

STOCK NAME	INDUSTRY	52 WEEK HIGH	MARKET PRICE
ABC	ENERGY	80	65
ABCD	CONSUMER	70	74
ABCDE	UTILITIES	96	82
ABCDEF	CHEMICALS	53	52
BDF	SERVICES	98	81
CDE	UTILITIES	78	37

Figure 11-4 Records in the COULD BUY table

For interpretation of the data in Figure 11-4, look at the record for stock BDF. The company is in the Services industry. Sometime in the past 52 weeks, the stock sold for $98 per share. However, now the stock sells for only $81 per share. Although Holly does follow this stock, she would not buy it at this time because $81 is more than 75% of $98.

To continue this case, you must have **INVEST.mdb**, which you can find with your Data files, in the Case 11 folder.

Assignment 1 Using Access and Excel for Decision Support

Holly has rules governing how she would sell and buy stocks shown in the database tables. You want to restructure the portfolio using the data in Access and Excel. You will use the following seven steps:

1. Create worksheets for the current portfolio and for the stocks that could be bought by importing Access table data.
2. In the worksheets, generate ratio data about stocks that should be bought and sold.
3. Augment the current portfolio worksheet to identify stocks that should be sold and to compute industry concentrations.
4. Prune rows from the stocks that could be bought to show data for only stocks that meet the market price to 52-week-high ratio requirement, sorted by desirability.
5. Create a worksheet of stocks retained from the current portfolio, pruning out those that should be sold.
6. Create a worksheet showing a new portfolio that combines retained stocks and stocks that could be bought.
7. Export the new portfolio data to Access.

Next, you will work through each of the seven steps.

Step 1: Create Stock Worksheets

You will create worksheets for the stocks that are owned and that could be bought. Be sure that the **INVEST.mdb** database is closed (imports into Excel will be difficult if the database file is open). Open an Excel worksheet. To start, save it as **INVEST.xls**.

Your spreadsheet should start with two worksheets: one that shows data about the stocks that are owned and one that shows data about the stocks that could be bought. These worksheets are discussed next:

- **CURRENT PORTFOLIO** worksheet
- **COULD BUY** worksheet

CURRENT PORTFOLIO Worksheet

To create this worksheet, import the Owned table into an Excel worksheet, presumably Sheet1 at this point. Follow this procedure:

1. Use the **Data** menu option.
2. Select **Import External Data** then select **Import Data**. Identify the **INVEST.mdb** file as the source file.
3. Select the **OWNED table** and then follow the prompts. The data will be brought into the worksheet.

The top few rows of the data are shown in Figure 11-5.

	A	B	C	D	E
1	STOCK NAME	INDUSTRY	SHARES	AVERAGE PRICE	MARKET PRICE
2	ACE	CAPITAL	1000	82	61
3	BCD	FINANCE	5000	76	36
4	DEF	TECHNOLOGY	2000	69	38
5	FGH	CAPITAL	5000	95	56
6	HIJ	HEALTHCARE	5000	86	63
7	JKL	MINING	1000	62	61

Figure 11-5 Data imported from the OWNED table

Renaming the worksheet is recommended for spreadsheet usability. (Excel's default sorting may result in an ordering different from that shown in Figure 11-2.) Rename the worksheet **CURRENT PORTFOLIO**, using this procedure:

1. To rename the worksheet, right-click the tab at the bottom of the worksheet.
2. Select **Rename** from the menu.
3. Type **CURRENT PORTFOLIO** in the tab.

The COULD BUY Worksheet

To create this worksheet, import the COULD BUY table into an Excel worksheet. The top few rows of the data are shown in Figure 11-6.

	A	B	C	D
1	STOCK NAME	INDUSTRY	52 WEEK HIGH	MARKET PRICE
2	ABC	ENERGY	80	65
3	CDE	UTILITIES	78	37
4	EFG	SERVICES	52	36
5	GHI	CONSUMER	92	45
6	IJK	CHEMICALS	45	78

Figure 11-6 Data imported from the COULD BUY table

Rename the worksheet **Could Buy**.

Step 2: Generate Ratio Data

Your next step is to augment the worksheets with the relevant "troubled stock" ratio data.

By formula, create the Sell Stock Ratio column in the **CURRENT PORTFOLIO** worksheet. Recall that this ratio's value will govern the selling decision. It is the ratio of the current market price to the average price. The first few rows of the worksheet should now look like the data shown in Figure 11-7.

	A	B	C	D	E	F
1	STOCK NAME	INDUSTRY	SHARES	AVERAGE PRICE	MARKET PRICE	SELL STOCK RATIO
2	ACE	CAPITAL	1000	82	61	0.744
3	BCD	FINANCE	5000	76	36	0.474
4	DEF	TECHNOLOGY	2000	69	38	0.551
5	FGH	CAPITAL	5000	95	56	0.589
6	HIJ	HEALTHCARE	5000	86	63	0.733
7	JKL	MINING	1000	62	61	0.984
8	LMN	FINANCE	5000	62	59	0.952

Figure 11-7 Data with the Sell Stock Ratio column

Conversely, the Buy Stock Ratio column should be put into the **COULD BUY** worksheet. Recall that this ratio's values will govern the buying decision. It is the ratio of the current market price to the 52-week-high price. The first few rows of the worksheet should now look like the data shown in Figure 11-8.

	A	B	C	D	E
1	STOCK NAME	INDUSTRY	52 WEEK HIGH	MARKET PRICE	BUY STOCK RATIO
2	ABC	ENERGY	80	65	0.813
3	CDE	UTILITIES	78	37	0.474
4	EFG	SERVICES	52	36	0.692
5	GHI	CONSUMER	92	45	0.489
6	IJK	CHEMICALS	45	78	1.733
7	KLM	FINANCE	94	63	0.670
8	MNO	TECHNOLOGY	64	50	0.781

Figure 11-8 Data with the Buy Stock Ratio

Step 3: Identify Stocks to Sell

Your next step should be to augment the worksheet with information needed for the Sell decision. Then you'll compute industry concentrations.

Use an IF() statement to identify those stocks that should be sold. The statement should implement this rule: Sell if the sell stock ratio is greater than or equal to 1.25; otherwise, keep the stock. The top part of the worksheet is shown in Figure 11-9. (All stocks shown in Figure 11-9 will be kept, but many others in the sheet will need to be sold.)

	A	B	C	D	E	F	G
1	STOCK NAME	INDUSTRY	SHARES	AVERAGE PRICE	MARKET PRICE	SELL STOCK RATIO	SELL?
2	ACE	CAPITAL	1000	82	61	0.744	KEEP
3	BCD	FINANCE	5000	76	36	0.474	KEEP
4	DEF	TECHNOLOGY	2000	69	38	0.551	KEEP
5	FGH	CAPITAL	5000	95	56	0.589	KEEP
6	HIJ	HEALTHCARE	5000	86	63	0.733	KEEP
7	JKL	MINING	1000	62	61	0.984	KEEP

Figure 11-9 Stocks to sell evaluated

Next, you should compute the sales value for stocks to be sold. If sold, the value is the number of shares times the current price. The sales value is zero if the stock will be kept. An IF() statement is needed. Continuing the example, Figure 11-10 shows the bottom part of the worksheet for the last stocks in the list, columns E through H. (Column headings are inserted above the figure to make things more clear; you would not see these when you scroll down in Excel.)

MARKET PRICE	SELL STOCK RATIO	SELL?	SELL FOR
E	F	G	H
97	1.198	KEEP	0
76	1.169	KEEP	0
111	1.542	SELL	555000
83	1.660	SELL	581000
86	1.049	KEEP	0
74	1.233	KEEP	0
105	1.382	SELL	945000

Figure 11-10 Sales value of stocks to sell

You should compute the book value (at the average price) of each stock and then use SUM() to compute the total sales value and total book values, respectively, as shown in Figure 11-11. In that figure, columns E through I are shown. (Again, column headings are inserted above the figure to make things more clear; you would not see these when you scroll down in Excel.)

MARKET PRICE	SELL STOCK RATIO	SELL?	SELL FOR	BOOK VALUE
E	F	G	H	I
97	1.198	KEEP	0	567000
76	1.169	KEEP	0	585000
111	1.542	SELL	555000	360000
83	1.660	SELL	581000	350000
86	1.049	KEEP	0	82000
74	1.233	KEEP	0	60000
105	1.382	SELL	945000	684000
			6279000	18513000

Figure 11-11 Total sales values and book values of stocks

Notice that the sales value of stocks that would be sold is about $6.28 million. Once sold, that $6.28 million would then be available to buy other stocks—from the stocks shown in the COULD BUY worksheet!

Finally, make a pivot table showing the total portfolio book value for each industry. Using the pivot table, compute the percentage that each industry bears to the whole portfolio. The output, shown in Figure 11-12, indicates that some industries have more than a 20% representation in the $18.5 million portfolio.

Sum of BOOK VALUE		INDUSTRY
INDUSTRY	Total	CONCENTRATION
CAPITAL	3223000	0.174
FINANCE	3414000	0.184
HEALTHCARE	3325000	0.180
MINING	4318000	0.233
TECHNOLOGY	4233000	0.229
Grand Total	18513000	1.000

Figure 11-12 Portfolio values and representation by industry

Step 4: Select Stocks

Next, you'll prune rows from the stocks that could be bought. Your goal is to show data for only those stocks that meet the market price to 52-week-high ratio requirement, sorted by desirability.

First, turn your attention to the **COULD BUY** worksheet. You want to indicate those stocks that could be bought under Holly's rule. Then, you'll weed out rows showing data for stocks that would not be bought.

In the first step, use an IF() statement to show which stocks to buy and which to ignore. If the buy stock ratio is less than or equal to .75, then Holly could buy the stock; otherwise, Holly would ignore it. The top rows in the results are shown in Figure 11-13.

	A	B	C	D	E	F
1	STOCK NAME	INDUSTRY	52 WEEK HIGH	MARKET PRICE	BUY STOCK RATIO	BUY?
2	ABC	ENERGY	80	65	0.813	IGNORE
3	CDE	UTILITIES	78	37	0.474	BUY
4	EFG	SERVICES	52	36	0.692	BUY
5	GHI	CONSUMER	92	45	0.489	BUY
6	IJK	CHEMICALS	45	78	1.733	IGNORE
7	KLM	FINANCE	94	63	0.670	BUY

Figure 11-13 Stocks to buy or ignore

The worksheet should then be edited to weed out rows of data for stocks that would be ignored. For such a stock, put the cursor in the left prefix area, which has the row number. Select Edit and then Delete to delete the row. After doing this repeatedly, the top part of the worksheet should look like that shown in Figure 11-14.

	A	B	C	D	E	F
1	STOCK NAME	INDUSTRY	52 WEEK HIGH	MARKET PRICE	BUY STOCK RATIO	BUY?
2	CDE	UTILITIES	78	37	0.474	BUY
3	EFG	SERVICES	52	36	0.692	BUY
4	GHI	CONSUMER	92	45	0.489	BUY
5	KLM	ENERGY	94	63	0.670	BUY
6	OPQ	SERVICES	91	39	0.429	BUY

Figure 11-14 Data for stocks to buy

It would be helpful to have this data sorted by ratio. Presumably, the stock with the lowest ratio is the most depressed in price and, therefore, the most desirable to buy. The Could Buy worksheet could be sorted, or a copy of the worksheet could be created and then sorted. The latter approach is recommended here:

1. Highlight the data range by selecting **Edit** and then **Copy**.
2. Select a blank worksheet. (If none exists, use **Insert—Worksheet** to make one.)
3. With your cursor in cell A1, select **Edit** and then **Paste**.
4. Rename the worksheet **COULD BUY SORTED**.
5. Data can be sorted by selecting the **Data** menu and then the **Sort** option. The Sort column is then identified—here the buy stock ratio value is the Sort field.
6. Select **Ascending** for low-to-high ordering.

The results are shown in Figure 11-15.

	A	B	C	D	E	F
1	STOCK NAME	INDUSTRY	52 WEEK HIGH	MARKET PRICE	BUY STOCK RATIO	BUY?
2	WXY	UTILITIES	89	35	0.393	BUY
3	KLMN	CONSUMER	93	38	0.409	BUY
4	OPQ	SERVICES	91	39	0.429	BUY
5	YZA	SERVICES	94	41	0.436	BUY
6	FHJ	CONSUMER	88	39	0.443	BUY
7	CDE	UTILITIES	78	37	0.474	BUY

Figure 11-15 Data about stocks to buy, sorted

Step 5: Select Stocks to Retain

You now want to create a worksheet that shows what will be retained from the current portfolio. The recommended procedure follows:

1. Make a copy of the **OWNED STOCKS** worksheet data.
2. Put the data into a new worksheet named **AFTER SELLING**.
3. In the **AFTER SELLING** worksheet, delete rows of data of stock to be sold.

The bottom part of the **AFTER SELLING** worksheet should look like that shown in Figure 11-16, columns A through I as before. (Column headings are inserted above the figure to make things more clear; you would not see these when you scroll down in Excel.)

1	STOCK NAME	INDUSTRY	SHARES	AVERAGE PRICE	MARKET PRICE	SELL STOCK RATIO	SELL?	SELL FOR	BOOK VALUE
	A	B	C	D	E	F	G	H	I
45	JKLMNO	HEALTHCARE	2000	100	102	1.020	KEEP	0	200000
46	LMNOPQ	MINING	8000	95	74	0.779	KEEP	0	760000
47	NOPQRS	FINANCE	7000	81	97	1.198	KEEP	0	567000
48	PQRSTU	TECHNOLOGY	9000	65	76	1.169	KEEP	0	585000
49	VWXYZA	MINING	1000	82	86	1.049	KEEP	0	82000
50	XYZABC	FINANCE	1000	60	74	1.233	KEEP	0	60000

Figure 11-16 Data about stocks to keep

Step 6: Create the New Portfolio

You can now turn your attention to constructing a worksheet showing the new portfolio that combines retained stocks and stocks that could be bought. Use the following procedure:

1. Make a copy of the **AFTER SELLING** worksheet data, leaving out the SELL STOCK RATIO and SELL? columns.
2. Name that worksheet the **NEW PORTFOLIO** worksheet.
3. The two right-most column headers should be for **Book Value** (shares times average price) and for **Market Value** (shares times current market price).

The top part of the **NEW PORTFOLIO** worksheet should then look like that shown in Figure 11-17.

	A	B	C	D	E	F	G
1	STOCK NAME	INDUSTRY	SHARES	AVERAGE PRICE	MARKET PRICE	BOOK VALUE	MARKET VALUE
2	ACE	CAPITAL	1000	82	61	82000	61000
3	BCD	FINANCE	5000	76	36	380000	180000
4	DEF	TECHNOLOGY	2000	69	38	138000	76000
5	FGH	CAPITAL	5000	95	56	475000	280000
6	HIJ	HEALTHCARE	5000	86	63	430000	315000
7	JKL	MINING	1000	62	61	62000	61000

Figure 11-17 Data about stocks to keep, including book values and market values

Insert a pivot table showing industry concentrations, as shown in Figure 11-18.

BOOK VALUE OF PORTFOLIO AFTER SELLING STOCKS		
Sum of BOOK VALUE		INDUSTRY
INDUSTRY	Total	CONCENTRATION
CAPITAL	2606000	0.178
FINANCE	2938000	0.201
HEALTHCARE	2420000	0.166
MINING	3634000	0.249
TECHNOLOGY	3010000	0.206
Grand Total	14608000	1.000

Figure 11-18 Value of stocks retained by industry concentration

Notice that some industries are over-represented by the 20% rule, so that stocks will have to be added to the list. Some rule will have to be followed when choosing these stocks. In the example here, an equal number of shares of each acceptable stock will be used, for illustration. Other stock purchase rules are possible. For example: stocks could be bought to cure industry under-representation. Or, stocks with the best ratios could be emphasized.

The data about stocks that could be bought should now be incorporated into the **NEW PORTFOLIO** worksheet.

The **COULD BUY SORTED** worksheet can now be used as a place to generate data about new stocks. Delete these columns: 52 Week High, Buy Stock Ratio, and Buy? The worksheet should then be modified to include column headers for Shares owned, Book Value, and Market Value. The result is a worksheet in the same format as the **NEW PORTFOLIO** worksheet, as shown in Figure 11-19.

	A	B	C	D	E	F	G
1	STOCK NAME	INDUSTRY	SHARES	AVERAGE PRICE	MARKET PRICE	BOOK VALUE	MARKET VALUE
2	WXY	UTILITIES	0	35	35	0	0
3	KLMN	CONSUMER	0	38	38	0	0

Figure 11-19 Format of revised COULD BUY SORTED worksheet

Looking at the columns in Figure 11-19, no shares are yet owned, so data values in the columns Book Value and Market Value are zero. At this point, the values in the column Average Price (if purchased) would be the same as those in the column for current Market Price. The bottom of the Book Value and Market Value columns should have formulas that total the columns.

You can then play "what-if" with the number of shares owned. You would do this until the total amount spent comes close to the total available. Continuing the example, about $6.28 million can be spent. (Assume that Holly can be a few dollars over or under that amount.) Also assume that Holly decides to buy an equal number of shares of each stock. By experimentation, this turns out to be 7,900 shares, as shown in Figure 11-20.

	A	B	C	D	E	F	G
1	STOCK NAME	INDUSTRY	SHARES	AVERAGE PRICE	MARKET PRICE	BOOK VALUE	MARKET VALUE
2	WXY	UTILITIES	7900	35	35	276500	276500
3	KLMN	CONSUMER	7900	38	38	300200	300200
4	OPQ	SERVICES	7900	39	39	308100	308100
5	YZA	SERVICES	7900	41	41	323900	323900
6	FHJ	CONSUMER	7900	39	39	308100	308100
7	CDE	UTILITIES	7900	37	37	292300	292300
8	GHI	CONSUMER	7900	45	45	355500	355500
9	CEG	CONSUMER	7900	47	47	371300	371300
10	NPQ	ENERGY	7900	48	48	379200	379200
11	QSV	UTILITIES	7900	49	49	387100	387100
12	GIK	CHEMICALS	7900	48	48	379200	379200
13	GHIJ	UTILITIES	7900	54	54	426600	426600
14	KLM	ENERGY	7900	63	63	497700	497700
15	EFGH	ENERGY	7900	56	56	442400	442400
16	EFG	SERVICES	7900	36	36	284400	284400
17	JLN	CHEMICALS	7900	47	47	371300	371300
18	CDEFGH	ENERGY	7900	74	74	584600	584600
19							
20						6288400	6288400

Figure 11-20 Purchase of an equal number of each stock

Other stock selection rules are possible, of course. For example, the user might opt to buy just the shares from under-represented industries. Or the user might opt to buy just the shares of the first five stocks—presumably, those with the lowest ratios will see the greatest market price increases in the future.

Once satisfied with the new stock selections, Holly could copy and then paste the data at the end of the **NEW PORTFOLIO** worksheet data. Continuing the example, the bottom part of that worksheet (showing columns A to G as before) would now look like that shown in Figure 11-21 (Book Value and Market Value columns have been totaled). (Column headings are inserted above the figure to make things more clear; you would not see these when you scroll down in Excel.)

1	STOCK NAME	INDUSTRY	SHARES	AVERAGE PRICE	MARKET PRICE	BOOK VALUE	MARKET VALUE
	A	B	C	D	E	F	G
61	GIK	CHEMICALS	7900	48	48	379200	379200
62	GHIJ	UTILITIES	7900	54	54	426600	426600
63	KLM	ENERGY	7900	63	63	497700	497700
64	EFGH	ENERGY	7900	56	56	442400	442400
65	EFG	SERVICES	7900	36	36	284400	284400
66	JLN	CHEMICALS	7900	47	47	371300	371300
67	CDEFGH	ENERGY	7900	74	74	584600	584600
68							
69						20896400	19678400

Figure 11-21 Portfolio with new stocks included

Figure 11-22 shows a pivot table with industry concentrations at this point.

BOOK VALUE OF PORTFOLIO AFTER BUYING NEW STOCK		
Sum of BOOK VALUE		INDUSTRY
INDUSTRY	Total	CONCENTRATION
CAPITAL	2606000	0.125
CHEMICALS	750500	0.036
CONSUMER	1335100	0.064
ENERGY	1903900	0.091
FINANCE	2938000	0.141
HEALTHCARE	2420000	0.116
MINING	3634000	0.174
SERVICES	916400	0.044
TECHNOLOGY	3010000	0.144
UTILITIES	1382500	0.066
Grand Total	20896400	1.000

Figure 11-22 Portfolio values and representation by industry concentration

The new portfolio does not fully conform to Holly's allocation rules: No industry has more than 20% of the value, but some industries have less than 8%. Thus, this portfolio would not be an acceptable option for Holly.

It might be that Holly would want to see data for more than one possible portfolio. At this point, the new portfolio data would be copied to another worksheet (which could be named **NEW PORTFOLIO-1**). Then, the **COULD BUY SORTED** worksheet would be used to identify another set of stocks to add. These data records would be added to the end of the **NEW PORTFOLIO** worksheet to create a second portfolio. The data for that portfolio would then be copied to another worksheet (named **NEW PORTFOLIO-2**). Pivot tables would be used to test Holly's allocation rules.

Step 7: Export the New Portfolio Data

After the makeup of the new portfolio has been established, the data can then be exported to Access in a form that conforms to the OWNED table in the **INVEST.mdb** database. To create data that will be sent to Access, this procedure could be followed:

1. **Create** a new worksheet called **EXPORT**.
2. **Copy** the new portfolio data to **EXPORT**.
3. **Delete extraneous data** so that only columns corresponding to OWNED table fields remain: Stock Name, Industry, Shares, Average Price, and Market Price.

Continuing the example here, the top of the **EXPORT** worksheet data would look like that shown in Figure 11-23.

	A	B	C	D	E
1	STOCK NAME	INDUSTRY	SHARES	AVERAGE PRICE	MARKET PRICE
2	ACE	CAPITAL	1000	82	61
3	BCD	FINANCE	5000	76	36
4	DEF	TECHNOLOG'	2000	69	38
5	FGH	CAPITAL	5000	95	56

Figure 11-23 EXPORT worksheet data values

The Excel spreadsheet should then be closed. The **INVEST.mdb** file would be opened. Assume you do not yet want to discard the old OWNED table and that you want Access to create a new table that will hold the **EXPORT** worksheet data. Follow these steps:

1. Start with the **Tables** object selected and **Create table by entering data** selected, as shown in Figure 11-24.

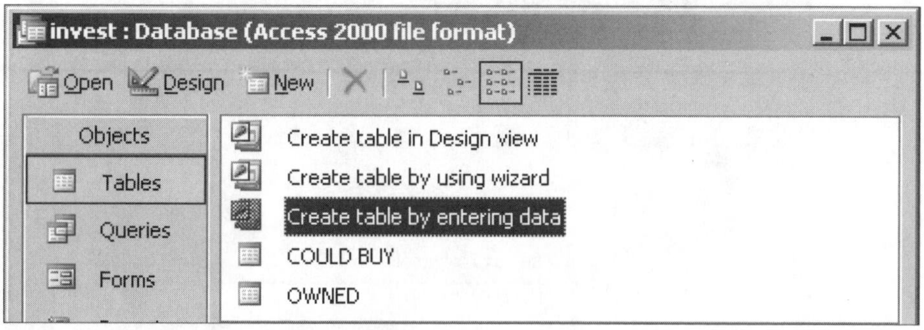

Figure 11-24 Beginning Import steps

2. Select **File—Get External Data** and then select **Import**. (You will then follow a series of prompts, not all of which are pictured here.)
3. Tell Access what file you want to use—**INVEST.xls**.
4. Use the **Import Spreadsheet Wizard** to tell Access to use the **EXPORT** worksheet, as shown in Figure 11-25.

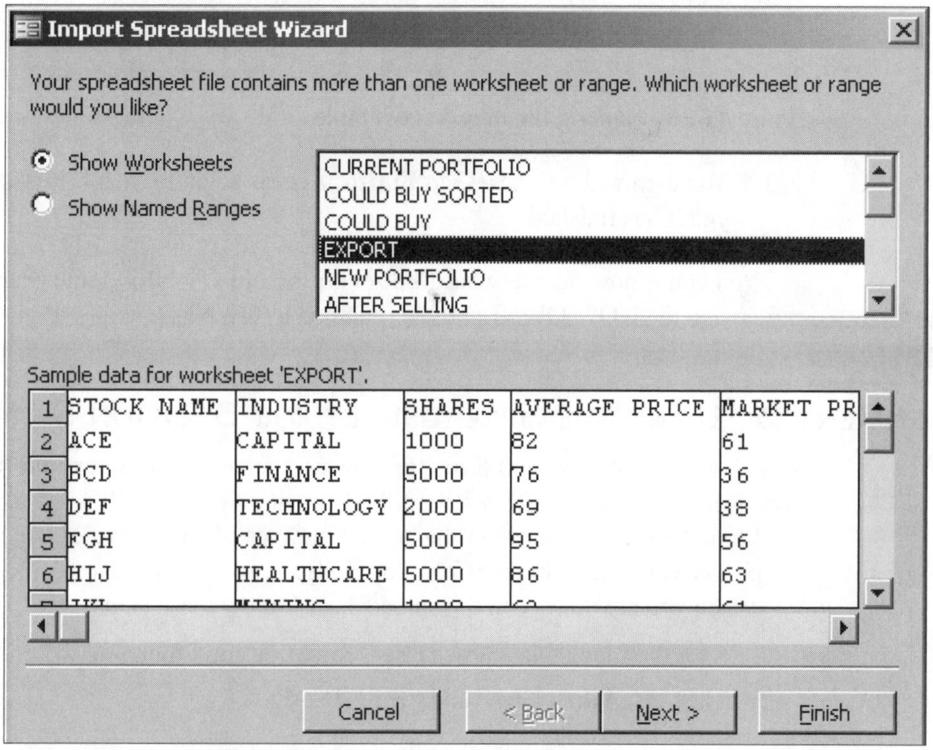

Figure 11-25 Identifying Excel worksheet in Import Spreadsheet Wizard

Case 11

5. Other prompts are then followed. At one point, you will tell Access what field to use as the primary key. Here, **Stock Name** would be the key, as it was in the OWNED table.
6. Eventually, you will tell Access what to name the new table. Assume that the table would be named NOW OWNED, as shown in Figure 11-26.

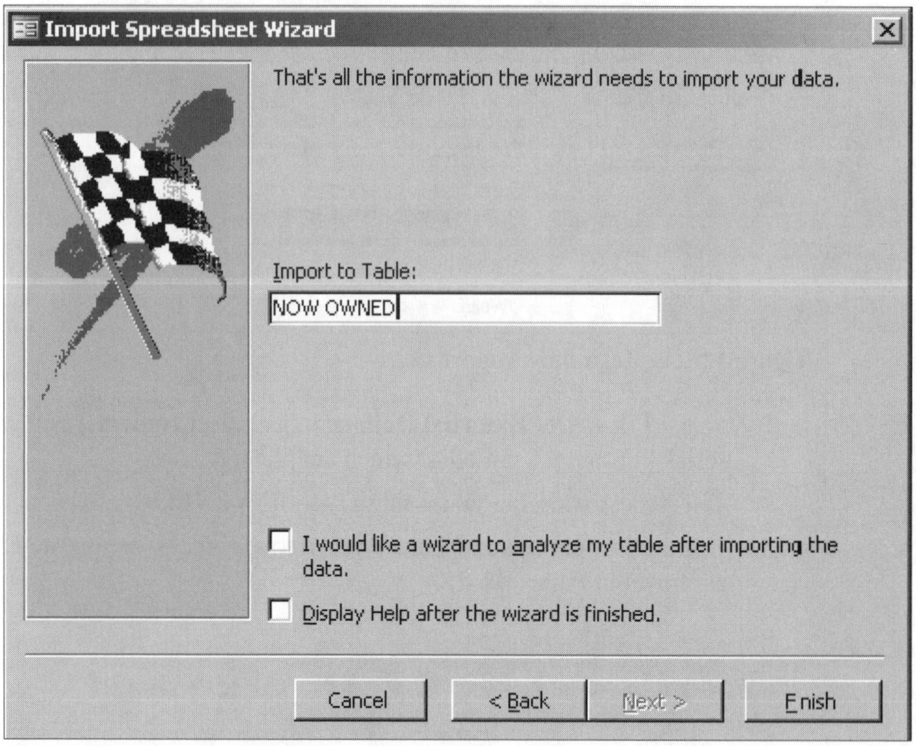

Figure 11-26 Naming the new Access table

7. You then will tell Access to **Finish**. Access actually then brings the data in and tells you it has finished.

You could now decide what to do with the old OWNED table. Perhaps it would be discarded and the NOW OWNED table renamed to OWNED.

ASSIGNMENT 2 USING THE SPREADSHEET FOR DECISION SUPPORT

You should use Access and Excel to construct an acceptable portfolio for Holly's consideration, following the procedures described previously. You will need to consistently follow a stock purchase rule other than the "equal number of shares" rule, which you now know will not produce an acceptable portfolio for Holly.

When you are done with the spreadsheet, follow these steps:

1. Save the file one last time **(File—Save)**. A good filename would be **INVEST.xls**.
2. Then, use **File—Close** and then **File—Exit**.

You are now in a position to document your work in a memorandum. Write a memorandum to Holly about your work. Observe the following requirements:

- Your memorandum should have a proper heading (DATE / TO / FROM / SUBJECT). You might want to use a Word memo template (**File**, **New**, click **On my computer** in the templates section, click the **Memos** tab, choose **Contemporary Memo**, and then click **OK**).
- Briefly outline the situation and your portfolio construction. Explain the stock purchase rule followed—cure under-representation, best ratio, or other rule.
- Support your memorandum graphically by inserting a summary table after the prose, like that shown in Figure 11-27.

Company	Number of Shares	Price	Value	% of Total
Company 1				
Company 2				
...				
Total				100%

Figure 11-27 Form for table in memorandum

Enter a table into Word, using the following procedure:

1. Select the **Table** menu option, point to **Insert**, then click **Table**.
2. Enter the number of rows and columns.
3. Select **AutoFormat** and choose **Table Grid 1**.
4. Select **OK**, and then select **OK** again.

Assignment 3 Giving an Oral Presentation

Assume that Holly is impressed by your "what-if" portfolio program, and thinks that other fund managers at her mutual fund company might want to know about it. She asks you to give a presentation explaining your program and your method. Prepare to explain your work and your recommendation to the group in 10 minutes or fewer. Use visual aids or handouts that you think are appropriate. Tutorial E has guidance on how to prepare and give an oral presentation.

Deliverables

1. Printouts of your memorandum
2. Disk or CD, which should have your Word memorandum file, your Excel spreadsheet *.xls* files, and your Access *.mdb* file.

Staple the printouts together, with the memorandum on top. If there are other *.xls* files or *.mdb* files on your disk, write your instructor a note, stating the names of the files pertinent to this case.

PART 6

Presentation Skills

Giving an Oral Presentation

Giving an oral presentation provides you with the opportunity to practice the presentation skills you'll need in the workplace. The presentations you'll create for the cases in this textbook will be similar to real-world presentations: You'll present objective, technical results to an organization's stakeholders. You'll support your presentation with visual aids commonly used in the business world. Your instructor might want to have your classmates role-play an audience of business managers, bankers, or employees and have them give you feedback on your presentation.

Follow these four steps to create an effective presentation:

1. Plan your presentation.
2. Draft your presentation.
3. Create graphics and other visual aids.
4. Practice your delivery.

Let's start at the beginning and look at the steps involved in planning your presentation.

Plan Your Presentation

When planning an oral presentation, you'll need to know your time limits, establish your purpose, analyze your audience, and gather information. Let's look at each of these elements.

Know Your Time Limits

You'll need to consider your time limits on two levels. First, consider how much time you'll have to deliver your presentation. What can you expect to accomplish in 10 minutes? The element of time is the "driver" of any presentation. It limits the breadth and depth of your talk—and the number of visual aids that you can use. Second, consider how much time you'll need for the actual process of preparing your presentation: drafting your presentation, creating graphics, and practicing your delivery.

Establish Your Purpose

You must define your purpose: what you need and want to say and to whom. For the cases in the Access portion of the book, your purpose will be to inform and explain. For instance, a business's owners, managers, and employees need to know how their organization's database is organized and how to use it to fill in input forms, create reports, and so on. By contrast, for the cases in the Excel portion of the book, your purpose will be to recommend a course of action. You'll be making recommendations based on your results from inputting various scenarios to business owners, managers, and bankers.

Analyze Your Audience

Before drafting your presentation, analyze your audience. Ask yourself these questions: What does my audience already know about the subject? What do they want to know? What do they need to know? Do they have any biases that I should consider? What level of technical detail is best suited to their level of knowledge and interest?

In some Access cases, you will make a presentation to an audience who might not be familiar with Access or databases in general. In other cases, you might be giving a presentation to a business owner who started work on the database but was not able to finish it. Tailor your presentation to suit your audience.

For the Excel cases, you will be interpreting results for an audience of bankers and business managers. The audience does not need to know the detailed technical aspects of how you generated your results. What they *do* need to know is what assumptions you made prior to developing your spreadsheet because those assumptions might have an impact on their opinion of your results.

Gather Information

Because you will have just completed a case, you'll have the basic information. For the Access cases, review the main points of the case and your goals. Be sure to include all the points that you think are important for the audience. In addition, you might want to go beyond the requirements and explain additional ways in which the database could be used to benefit the organization, now or in the future.

For the Excel cases, you can refer to the tutorials for assistance in interpreting the results from your spreadsheet analysis. For some cases, you might want to research the Internet for business trends or background information that could be used to support your presentation.

DRAFTING YOUR PRESENTATION

You might be tempted to write your presentation and then memorize it, word for word. If you do, your presentation will sound very unnatural because when people speak, they use a simpler vocabulary and shorter sentences than when they write. Thus, you might want to draft your presentation by noting just key phrases and statistics. When drafting your presentation, follow this sequence:

1. Write the main body of your presentation.
2. Write the introduction to your presentation.
3. Write the conclusion to your presentation.

Writing the Main Body

When you draft your presentation, write the body first. If you try to write the opening paragraph first, you'll spend an inordinate amount of time creating "the perfect paragraph"—only to revise it after you've written the body of your presentation.

Keep Your Audience in Mind

To write the main body, review your purpose and your audience's profile. What are the main points you need to make? What are your audience's wants, needs, interests, and technical expertise? It's important to include some basic technical details in your presentation, but keep in mind the technical expertise of your audience.

What if your audience consists of people with different needs, interests, and levels of technical expertise? For example, in the Access cases, an employee might want to know how to input information into a form, but the business owner might already know how to input data and may be more interested in generating queries and reports. You'll need to acknowledge their differences in your presentation. Thus, you might want to say something like, "And now, let's look at how data entry clerks can input data into the form."

Similarly, in the Excel cases, your audience will usually consist of business owners, managers, and bankers. The owners' and managers' concerns will be profitability and growth. By contrast, the bankers' main concern will be getting a loan repaid. You'll need to address the interests of each group.

Use Transitions and Repetition

Because your audience can't read the text of your presentation, you'll need to use transitions to compensate. Words such as *next*, *first*, *second*, and *finally* will help your audience follow the sequence of your ideas. Words such as *however*, *by contrast*, *on the other hand*, and *similarly* will help them follow shifts in thought. You can also use your voice and hand gestures to convey emphasis.

Also think about how you can use body language to emphasize what you're saying. For instance, if you are stating three reasons, you can tick them off on your fingers as you discuss them: one, two, three. Similarly, if you're saying that profits will be flat, you can make a level motion with your hand for emphasis.

As you draft your presentation, repeat key points to emphasize them. For example, suppose that your point is that outsourcing labor will provide the greatest gains in net income. Begin by previewing that concept: State that you're going to demonstrate how outsourcing labor will yield the greatest profits. Then provide statistics that support your claim and show visual aids that graphically illustrate your point. Summarize by repeating your point: "As you can see, outsourcing labor does yield the greatest profits."

Rely on Graphics to Support Your Talk

As you write the main body, think of how you can best incorporate graphics into your presentation. Don't waste a lot of words describing what you're presenting if you can use a graphic that can quickly portray it. For instance, instead of describing how information from a query is input into a report, show a sample, a query result, and a completed report. Figures E-1 and E-2 illustrate this.

Query for Report : Select Query

Last Name	First Name	Date	Type	Amount Paid
Angerstein	Amy	6/30/2006	Catalina 23	20
Angerstein	Amy	7/2/2006	Catalina 23	35
Codfish	Colburn	7/5/2006	Capri 14.2	60
Dellabella	Debra	7/5/2006	Hobie 18	20
Edelweiss	Edna	7/11/2006	Sunfish 12	150
Angerstein	Amy	8/1/2006	Row	30

Figure E-1 Access query

Summary Report

Last Name	*First Name*	*Date*	*Type*	*Amount Paid*
Angerstein	*Amy*			
		8/1/2006	Row	$30.00
		7/2/2006	Catalina 23	$35.00
		6/30/2006	Catalina 23	$20.00
	Total			*$85.00*
Codfish	*Colburn*			
		7/5/2006	Capri 14.2	$60.00
	Total			*$60.00*

Figure E-2 Access report

Also consider what kinds of graphics media are available—and how well you know how to use them. For example, if you've never prepared a PowerPoint presentation, will you have enough time to learn how to do it before your presentation?

Anticipate the Unexpected

Even though you're just drafting your report now, eventually you'll be answering audience questions. Being able to handle questions smoothly is the mark of a professional. The first step is being prepared for those questions.

You won't use all the facts you gather in your presentation. However, as you draft your presentation, you might want to keep some of those facts jotted on paper—just in case you need them to answer questions from the audience. For instance, for some Excel presentations you might be asked why you are not recommending some course of action that you did not mention in your report.

Writing the Introduction

After you have written the main body of your talk, then develop an introduction. An introduction should be only a paragraph or two in length and preview the main points that your presentation will cover.

For some of the Access cases, you might want to include some general information about databases: what they can do, why they are used, and how they can help the company become more efficient and profitable. You won't need to say much about the business operation because the audience already works for the company.

For the Excel cases, you might want to have an introduction to the general business scenario and describe any assumptions you made when creating and running your decision support spreadsheet. Excel is used for decision support, so describe the choices and decision criteria.

Writing the Conclusion

Every good presentation needs a good ending. Don't leave the audience hanging! Your conclusion should be brief—only a paragraph or two in length—and give your presentation a sense of closure. Use the conclusion to repeat your main points or, for the Excel cases, your findings and/or recommendations.

Creating Graphics

Using visual aids is a powerful method of getting your point across and making it understandable to your audience. Visual aids come in a variety of physical forms. Some forms are more effective than others.

Choosing Graphics Media

The media that you use should depend on your situation and what media are available. One of the key things to remember when using any media is this: *You must maintain control of the media or you'll lose control of your audience.*

The following list highlights some of the most common media and their strengths and weaknesses.

- **Handouts:** This medium is readily available in both classrooms and businesses. It relieves the audience from taking notes. Graphics can be in full color, of professional quality, and multicolored. *Negatives*: You must stop and take time to hand out individual sheets. During your presentation, the audience might be studying and discussing your handouts rather than listening to you. Lack of media control is *the* major drawback—and it can kill your presentation.
- **Chalkboard:** This informal medium is readily available in the classroom but not in many businesses. *Negatives*: You'll need to turn your back to the audience when you're writing (thus losing control of them), and you'll need to erase what you've written as you go. Your handwriting must be very good. In addition, attractive graphics are difficult to create.
- **Flip Chart:** This informal medium is readily available in many businesses. *Negatives*: The writing space is so small, it's not effective for more than a very small group. This medium shares many of the same negatives as the chalkboard.
- **Overheads:** This medium is readily available both in classrooms and in businesses. You do have control over what the audience sees and when. You can create very professional PowerPoint presentations on overhead transparencies. *Negatives*: Handwritten overheads look amateurish. Without special equipment and preparation, graphics are difficult to do well.
- **Slides:** This formal medium is readily available in many businesses and can be used in large rooms. Slides can be either 35mm or the more popular electronic on-screen slides, which are usually *the* medium of choice for large organizations and are generally preferred for formal presentations. *Negatives*: You must have access to the equipment

and know how to use it. It takes time to learn how to create and use computer graphics. Also, you must have some source of ambient light, or it will be difficult to see your notes in the dark.

Creating Charts and Graphs

Technically, charts and graphs are not the same thing, although many graphs are called "charts." Usually, charts show relationships, and graphs show change. However, Excel makes no distinction and calls both charts.

Charts are easy to create in Excel. Unfortunately, they are so easy to create that people often create graphics that are meaningless or that inaccurately reflect the data they represent. Let's look at how to select the most appropriate graphics.

Charts

Use pie charts to display data that is related to a whole. Excel takes the numbers you want to graph and makes them a percentage of 100. You might use a pie chart when showing the percentage of shoppers who bought a generic brand of toothpaste versus a major brand, as shown in Figure E-3. You would *not*, however, use a pie chart to show a company's net income over a three-year period, because the period cannot be considered "a whole," or the years its "parts," as shown in Figure E-4.

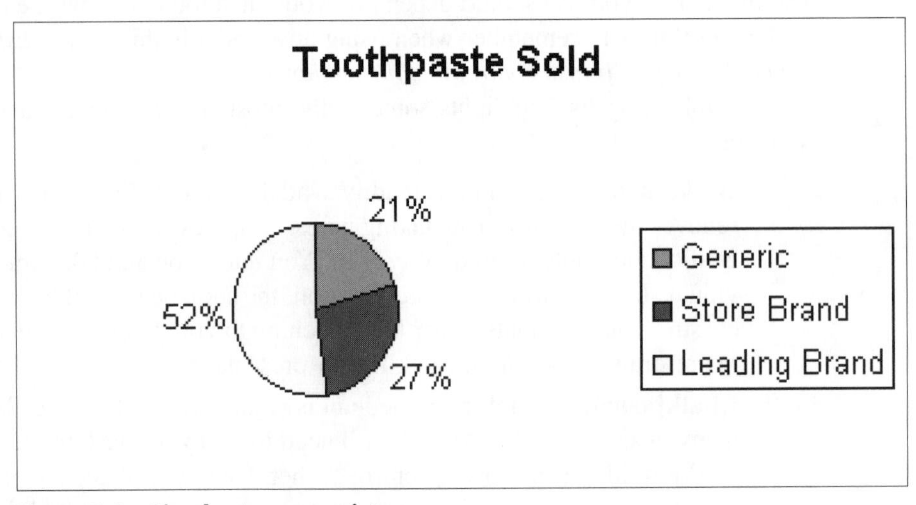

Figure E-3 Pie chart: appropriate use

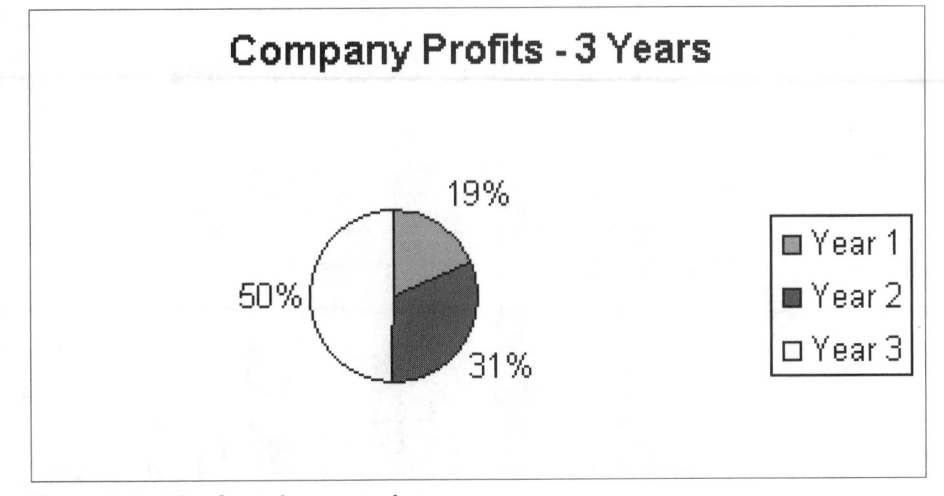

Figure E-4 Pie chart: inappropriate use

Use bar charts when you want to compare several amounts at one time. For example, you might want to compare the net profit that would result from each of several different strategies. You can also use a bar chart to show changes over time. For example, you might show how one pricing strategy would increase profits year after year.

When you are showing a graphic, don't forget that you need labels that explain what the graphic shows. For instance, if you're showing a graph with an X and Y axis, you should show what each axis represents so the audience doesn't puzzle over the graphic while you're speaking. Figures E-5 and E-6 show the necessity of labels.

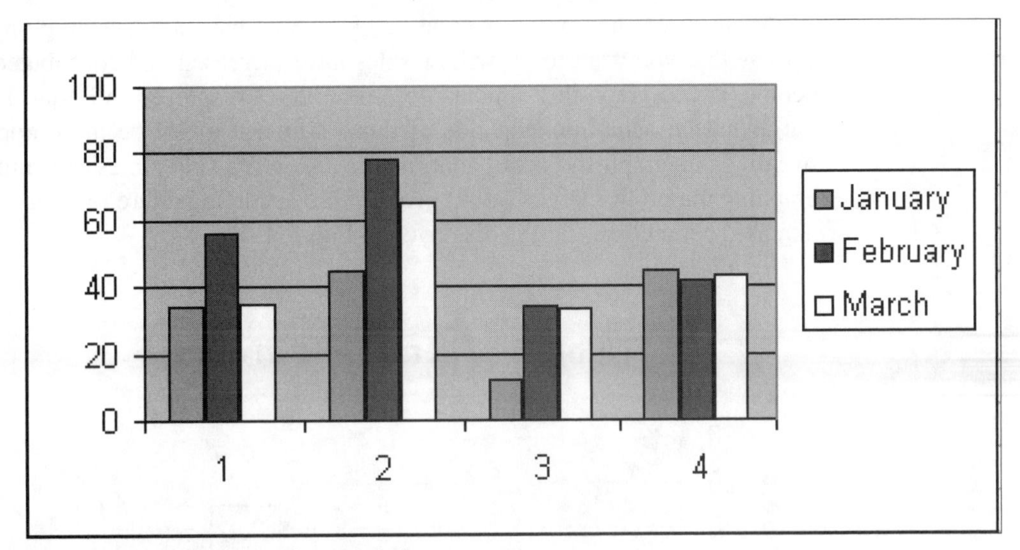

Figure E-5 Graphic without labels

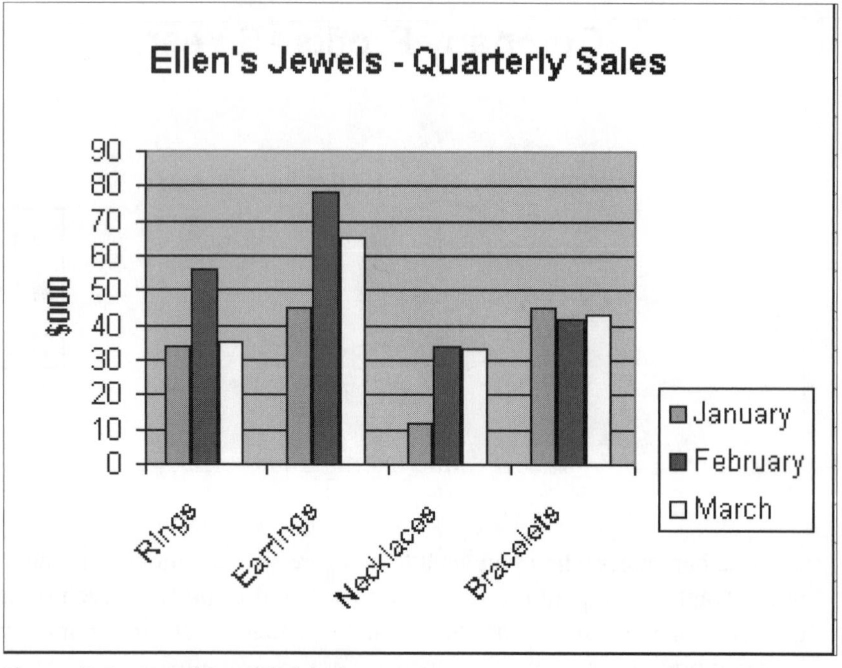

Figure E-6 Graphic with labels

In Figure E-5, the graphic is not labeled, and neither are the X and Y axes: Are the amounts shown units or dollars? What elements are represented by each bar? By contrast, Figure E-6 provides a comprehensive snapshot of the business operation—which would support a talk rather than distract from it.

Another common pitfall is creating charts that have a misleading premise. For example, suppose that you want to show how sales have increased and contributed to a growth in net income. If you graph the number of items sold, as displayed in Figure E-7, it might not tell you about the actual dollar value of those items; it might be more appropriate (and more revealing) to graph the profit margin for the items sold times the number of items sold. Graphing the profit margin would give a more accurate picture of what is contributing to the increased net income. This is displayed in Figure E-8.

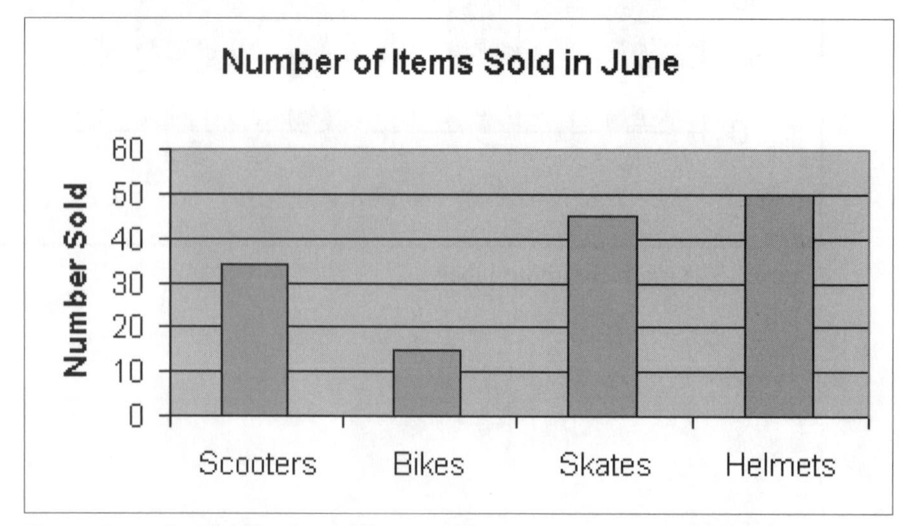

Figure E-7 Graph: number of items sold

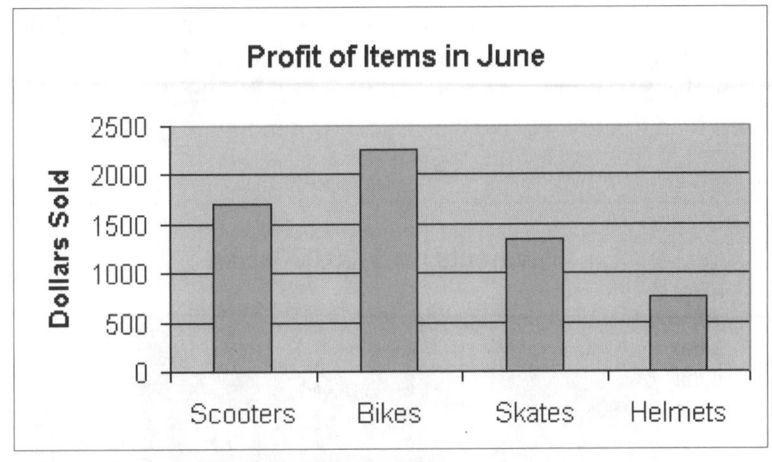

Figure E-8 Graph: profit of items sold

Another common pitfall is putting too much data in a single, comparative chart. Here is an example: Assume that you want to compare monthly mortgage payments for two loans with different interest rates and timeframes. You have a spreadsheet that computes the payment data, shown in Figure E-9.

Calculation of Monthly Payment						
Rate	6.00%	6.10%	6.20%	6.30%	6.40%	6.50%
Amount	100000	100000	100000	100000	100000	100000
Payment (360 payments)	$599	$605	$612	$618	$625	$632
Payment (180 payments)	$843	$849	$854	$860	$865	$871
Amount	150000	150000	150000	150000	150000	150000
Payment (360 payments)	$899	$908	$918	$928	$938	$948
Payment (180 payments)	$1,265	$1,273	$1,282	$1,290	$1,298	$1,306

Figure E-9 Calculation of monthly payment

You try to capture all this information in a single Excel chart, such as the one shown in Figure E-10.

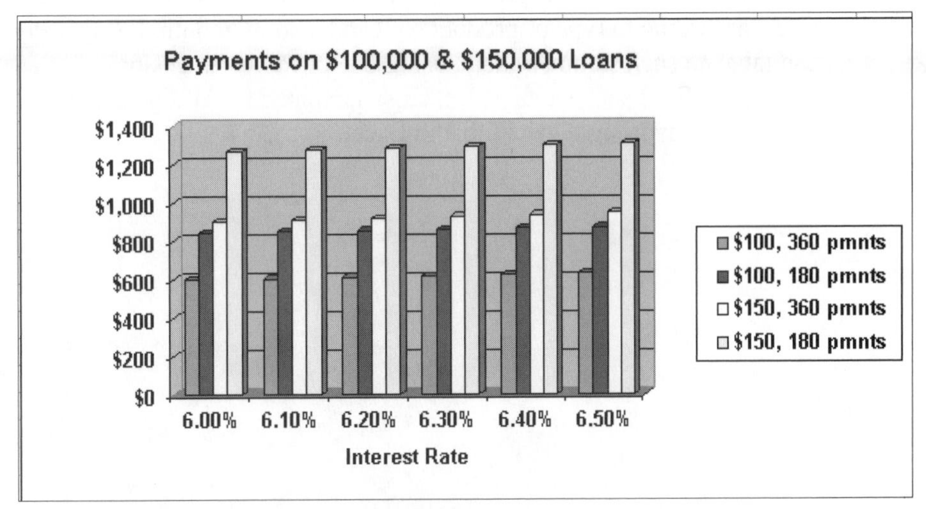

Figure E-10 Too much information in one chart

There is a great deal of information here. Most readers would probably appreciate it if you broke things up a bit. It would probably be easier to understand the data if you made one chart for the $100,000 loan and another one for the $150,000 loan. The chart for the $100,000 loan would look like the chart shown in Figure E-11.

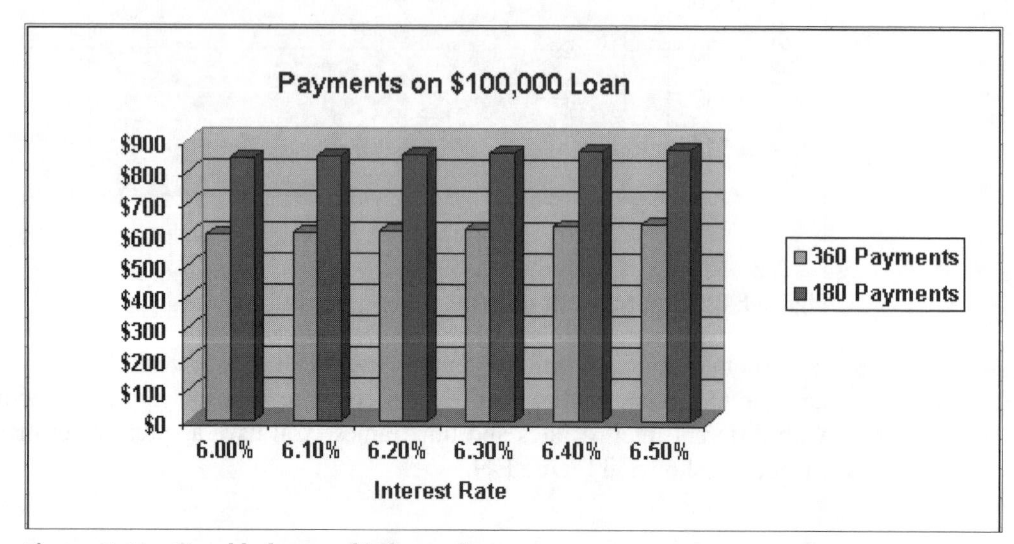

Figure E-11 Good balance of information

A similar chart could be made for the $150,000 loan. The charts could then be augmented by text that summarizes the main differences between the payments for each loan. In this fashion, the reader is led step by step through the data analysis.

You might want to use the Chart Wizard in Excel, but be aware that the charting functions can be tricky to use at times, especially with sophisticated charting. Some tweaking of the chart is often necessary. Your instructor might be able to provide specific directions for your individual charts.

Making a PivotTable in Excel

Suppose that you have data for a company's sales transactions by month, by salesperson, and by amount for each product type. You would like to display each salesperson's total sales, according to type of product sold and also by month. Using a PivotTable in Excel, you can tabulate such summary data, using one or more dimensions.

Figure E-12 shows total sales, cross-tabulated by salesperson and by month. This display was created by using a PivotTable in Excel.

	A	B	C	D	E
1	**Name**	**Product**	**January**	**February**	**March**
2	Jones	Product 1	30,000	35,000	40,000
3	Jones	Product 2	33,000	34,000	45,000
4	Jones	Product 3	24,000	30,000	42,000
5	Smith	Product 1	40,000	38,000	36,000
6	Smith	Product 2	41,000	37,000	38,000
7	Smith	Product 3	39,000	50,000	33,000
8	Bonds	Product 1	25,000	26,000	25,000
9	Bonds	Product 2	22,000	25,000	24,000
10	Bonds	Product 3	19,000	20,000	19,000
11	Ruth	Product 1	44,000	42,000	33,000
12	Ruth	Product 2	45,000	40,000	30,000
13	Ruth	Product 3	50,000	52,000	35,000

Figure E-12 Excel spreadsheet data

You can create this kind of table (and many other kinds) with the Excel PivotTable tool. You can use the following steps to create a PivotTable.

1. Select Data—PivotTable and PivotChart Report. You will see the screen shown in Figure E-13.

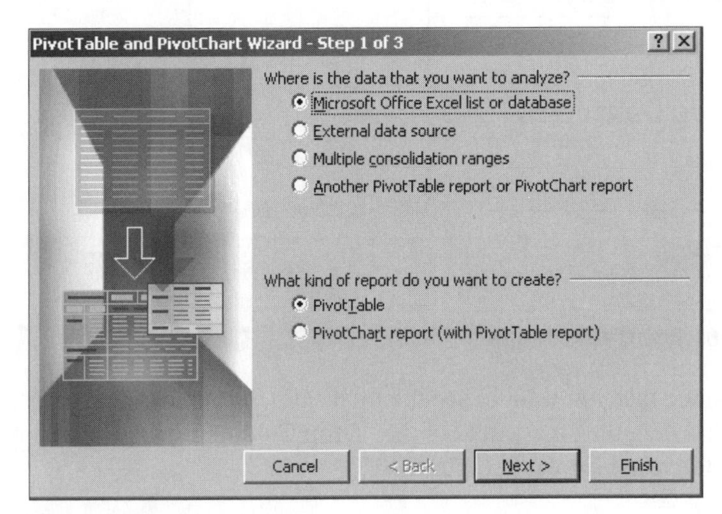

Figure E-13 Step 1

2. To make a PivotTable, click Next. You will see the screen shown in Figure E-14. By default the range will be the most northeast contiguous data range in the spreadsheet. You can change this in the Range window.

PivotTable and PivotChart Wizard - Step 2 of 3

Where is the data that you want to use?

Range: Sheet1!$A$1:$E$13 Browse...

Cancel < Back Next > Finish

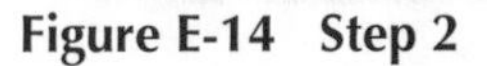

Figure E-14 Step 2

3. Click Next. You will then see the screen shown in Figure E-15.

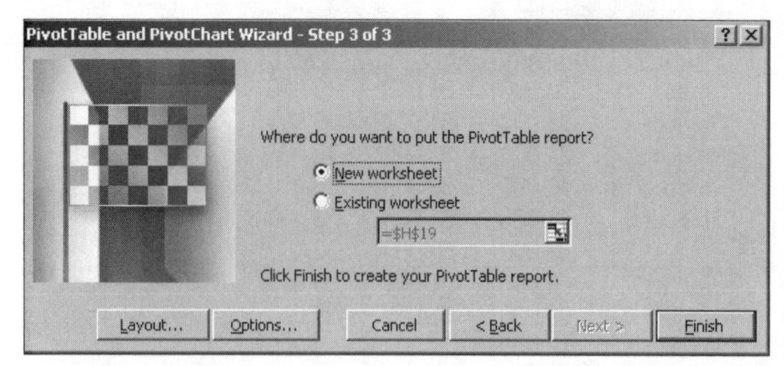

Figure E-15 Step 3

4. You can put the table in the current sheet (probably "Sheet1") or in a separate sheet. The latter way is shown here. *New worksheet* is the default. Click Finish.
5. You will see the screen shown in Figure E-16. The data range's column headings are shown in the PivotTable Field List. Click and drag column headings into the Row, Column, and Data areas.

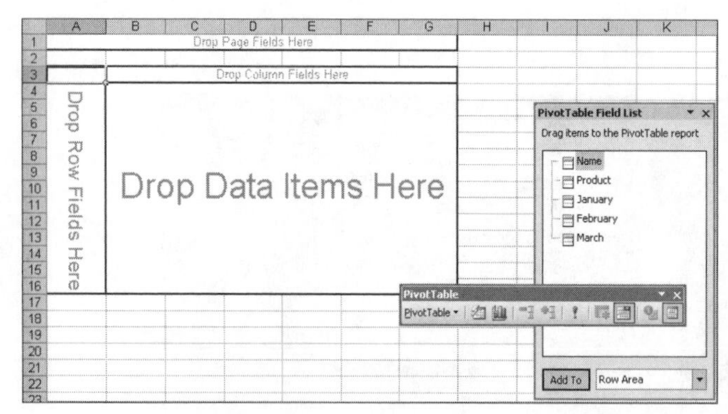

Figure E-16 PivotTable design

6. Assume that you want to see the total sales, by product, for each salesperson. You would drag the Name field to the "Drop Column Fields Here" area, and you should see the result shown in Figure E-17.

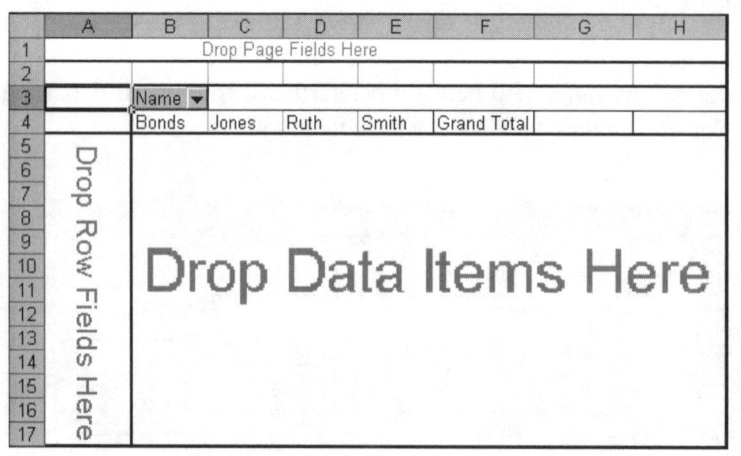

Figure E-17 Column fields

7. Next, take the Product field and drag it to the "Drop Row Fields Here" area, and you should see the result shown in Figure E-18.

	A	B	C	D	E	F
1	Drop Page Fields Here					
2						
3		Name				
4	Product	Bonds	Jones	Ruth	Smith	Grand Total
5	Product 1	Drop Data Items Here				
6	Product 2					
7	Product 3					
8	Grand Total					

Figure E-18 Row fields

8. Finally, take the month fields (January, February, and March) and drag them individually to the "Drop Data Items Here" area to produce the final PivotTable; you should see the result shown in Figure E-19.

	A	B	C	D	E	F	G
1							
2							
3			Name				
4	Product	Data	Bonds	Jones	Ruth	Smith	Grand Total
5	Product 1	Sum of January	25000	30000	44000	40000	139000
6		Sum of February	26000	35000	42000	38000	141000
7		Sum of March	25000	40000	33000	36000	134000
8	Product 2	Sum of January	22000	33000	45000	41000	141000
9		Sum of February	25000	34000	40000	37000	136000
10		Sum of March	24000	45000	30000	38000	137000
11	Product 3	Sum of January	19000	24000	50000	39000	132000
12		Sum of February	20000	30000	52000	50000	152000
13		Sum of March	19000	42000	35000	33000	129000
14	Total Sum of January		66000	87000	139000	120000	412000
15	Total Sum of February		71000	99000	134000	125000	429000
16	Total Sum of March		68000	127000	98000	107000	400000

Figure E-19 Data items

By default, Excel adds up all the sales for each salesperson by month for each individual product. It also shows the total sales for each month for all products at the bottom of the PivotTable.

Creating PowerPoint Presentations

PowerPoint presentations are easy to create: Simply open up the application and use the appropriate slide layout for a title slide, a slide containing a bulleted list, a picture, a graphic, and so on. In choosing a design template (the background color, the font color and size, and the fill-in colors for all slides in your presentation), keep these guidelines in mind:

- Avoid using pastel background colors. Dark backgrounds such as blue, black, and purple work well on overhead projection systems.

- If your projection area is small or your audience is large, you might want to use boldface type for all your text to make it even more visible.
- Try using "transition" slides to keep your talk lively. A variety of styles are in the program and available for use. Common transitions include "dissolves" and "wipes." Avoid wild transitions, such as swirling letters, that will distract your audience from your presentation.
- You can use "build" effects if you do not want your audience to see the whole slide when you show it. A "build" effect will allow each bullet to come up when the mouse button or the right arrow is clicked. A "build" effect allows you to control the visual and explain the elements as you go. This can be controlled under the Custom Animation screen, as shown in Figure E-20.

Figure E-20 Custom Animation screen

- You can create PowerPoint slides that have a section for notes. These are printed for the speaker when you choose Notes Pages from the *Print what* drop-down menu on the Print dialog box, as shown in Figure E-21. Each slide is printed as half-size, with the notes written underneath each slide, as shown in Figure E-22.
- As previously mentioned, always check your presentation on the overhead. What looks good on your computer screen might not be readable on an overhead screen.

Print

Printer

Name: \\prnserv\2100N_P225 | Properties

Status: Idle | Find Printer...

Type: HP LaserJet 2100 Series PS

Where: IP_128.175.21.6

Comment: Drivers updated on September 21, 2002 | Print to file

Print range

All | Current slide | Selection

Custom Show:

Slides:

Enter slide numbers and/or slide ranges. For example, 1,3,5-12

Copies

Number of copies: 1

Collate

Print what: Notes Pages

Color/grayscale: Grayscale

Handouts

Slides per page: 6

Order: Horizontal | Vertical

Scale to fit paper | Print hidden slides

Frame slides

Include comment pages

Preview | OK | Cancel

Figure E-21 Printing notes page

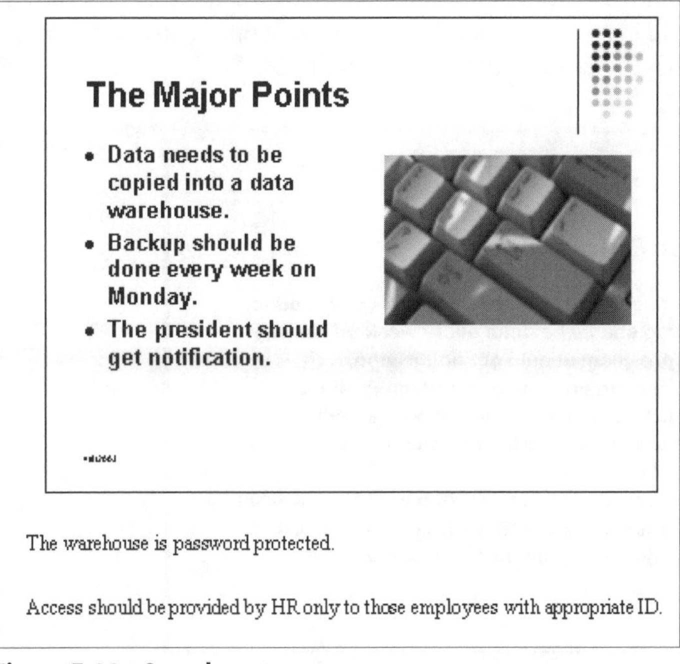

Figure E-22 Sample notes page

Tutorial E

Using Visual Aids Effectively

Make sure that you've chosen visual aids that will work for you most effectively. Also make sure that you have enough—but not too many—visual aids. How many is too many? The amount of time you have to speak will determine the number of visual aids that you should use, and so will your audience. For example, if you will be addressing a group of teenage summer helpers, you might want to use more visual effects than if you make a presentation to a board of directors. Remember, use visual aids to enhance your talk, not replace it.

Review each visual aid you've created to make sure that it meets the following criteria:

- The size of the visual aid is large enough so that everyone in the audience can see it clearly and read any labels.
- The visual aid is accurate, for example, the graphics are not misleading and there are no typos or misspelled words.
- The content of the visual aid is relevant to the key points of your presentation.
- The visual aid doesn't distract the audience from your message. Often when creating PowerPoint slides, speakers get carried away with the visual effects, for example, they use spiraling text and other jarring effects. Keep it professional.
- A visual aid should look good in the presentation environment. If at all possible, try using your visual aid in the presentation environment. For example, when using PowerPoint, try it out on the overhead projector and in the room in which you'll be showing the slides. What looks good on your computer screen might not look good on the overhead projector when viewed from a distance of 20 feet.
- Make sure that all numbers are rounded unless decimals or pennies are crucial.
- Do not make your slides too busy or crowded. Most experts say that bulleted lists should contain no more than four or five lines. Also avoid having too many labels. A busy slide is illustrated in Figure E-23.

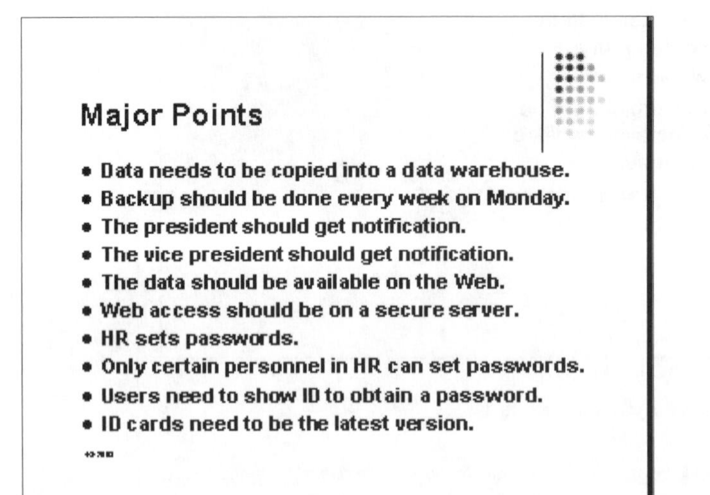

Figure E-23 Busy slide

PRACTICING FOR YOUR DELIVERY

Surveys indicate that public speaking is most people's greatest fear. However, fear or nervousness can be a positive factor. It can channel your energy into doing a good job. Remember that an audience will rarely perceive that you are nervous unless you fidget or your voice cracks. They are present to hear the content of your talk, so think of the audience, not how you feel.

The presentations you give for the cases in this textbook will be in a classroom setting with 20 to 40 students. Ask yourself this question: Would I be afraid to talk to just one or two of my classmates? Think of your presentation as an extended conversation with several of your classmates. Let your gaze shift from person to person and make eye contact with them. As your gaze drifts around the room, say to yourself, "I'm speaking to one person." As you become more experienced in speaking before groups, you will be able to let your gaze move naturally from one audience member to another.

Tips for Practicing Your Delivery

Giving an effective presentation is not reading a report to an audience. Rather, it requires that you have your message rehearsed well enough so you can present it naturally, confidently, and in tandem with well-chosen visual aids. Make sure that you allow sufficient time to practice your delivery.

- Practice your presentation several times and use your visual aids when you practice.
- Show visual aids at the right time and only at the right time. A visual aid should not be shown too soon or too late. In your speaker's notes, you might even have cues for when to show each visual aid.
- Maintain eye and voice contact with the audience when using the visual aid. Don't look at the screen or turn your back on the audience.
- Use your visual aids and refer to them both in your talk and with hand gestures. Don't ignore your own visual aid.
- Keep in mind that your visual aids should support your presentation, not *be* the presentation. In other words, don't have everything you are going to say on each slide. Use visual aids to illustrate the key points and statistics and fill in with your talk.
- Time check: Are you within time limits?
- Using numbers effectively: Use round numbers when speaking or you'll sound like a computer. Also, make numbers as meaningful as possible: For example, instead of saying "in 84.7 percent of cases," say, "in five out of six cases."
- Don't "reach" to interpret the output of statistical modeling. For example, suppose that you have input many variables into an Excel model. You might be able to point out a trend, but you might not be able to say with certainty that if management employs the inputs in the same combination that you used them, they will get exactly the same results.
- Record yourself, if possible, and then evaluate yourself. If that is not possible, have a friend listen to you and evaluate your style. Are you speaking down to your audience? Is your voice unnaturally high-pitched from fear? Are you speaking clearly and distinctly? Is your voice free of distractions, such as "um" and "you know," "uh, so," and "well"?

- If you use a pointer, either a laser pointer or a wand, use it with care. Make sure that you don't accidentally point a laser pointer in someone's face—you'll temporarily blind them. If you're using a wand, don't swing it around or play with it.

Handling Questions

Fielding questions from an audience can be an unpredictable experience because you can't anticipate all the questions that might be asked. When answering questions from an audience, *always treat everyone with courtesy and respect*, no matter what. Use the following strategies to handle questions:

- Anticipate questions. You can gather much of the information that you need as you draft your presentation. Also, if you have a slide that illustrates a key point but doesn't quite fit in your talk, save it—someone might have a question that the slide will answer.
- Mention at the beginning of the talk that you will take questions at the end of your talk. This will (you hope) prevent people from interrupting your presentation. If someone tries to interrupt you, smile and say that you'll be happy to answer all questions when you're finished or that the next graphic will answer their question. (If, however, the person doing the interrupting is the CEO of your company, you want to stop your presentation and answer the question on the spot.)
- When answering a question, first repeat the question if you have *any* doubt that the entire audience might not have heard it. Then deliver the answer to the whole audience, not just the one person who asked the question.
- Be informative and not persuasive, that is, use facts to answer questions. For instance, if someone asks your opinion about some outcome, you might show an Excel slide that displays the Solver's output, and then you can use that data as the basis for answering the question.
- If you don't know the answer to a question, don't try to fake it. For instance, suppose someone asks you a question about the Scenario Manager that you just can't answer. Be honest. Say, "That is an excellent question but, unfortunately, it's not one that I'm able to answer." At that point, you might ask your instructor whether he or she can answer the question. In a professional setting, you might say that you'll research the answer and e-mail the answer to the person who asked the question.
- Signal when you are finished. You might say, "I have time for one more question." Wrap up the talk yourself.

Handling a "Problem" Audience

A "problem" audience or a heckler is every presenter's nightmare. Fortunately, such experiences are rare. If someone is rude to you or challenges you in a hostile manner, keep cool, be professional, and rely on facts. Know that the rest of the audience sympathizes with your plight and admires your self-control.

The problem that you will most likely encounter is a question from an audience member who lacks technical expertise. For instance, suppose that you explained how to input data into an Access form, but someone didn't understand the explanation that you gave. In such an instance, ask the questioner what part of the explanation is confusing. If you can answer the question briefly, do so. If your answer to the questioner begins to turn into a time-consuming dialogue, offer to give the person one-on-one input later.

Another common problem is someone who asks you a question that you've already answered. The best solution is to answer the question as briefly as possible and use different words (just in case it's the way in which you explained something that confused the person). If the person persists in asking questions that have very obvious answers, either the person is clueless or is trying to heckle you. In that case, you might ask the audience, "Who in the audience would like to answer that question?" The person asking the question will get the hint.

Presentation Toolkit

You can use these forms for preparation, self-analysis, and evaluation of your classmates' presentations (Figures E-24, E-25, and E-26).

Preparation Checklist

Facilities and Equipment

❑ The room contains the equipment that I need.
❑ The equipment works and I've tested it with my visual aids.
❑ Outlets and electrical cords are available and sufficient.
❑ All the chairs are aligned so that everyone can see me and hear me.
❑ Everyone will be able to see my visual aids.
❑ The lights can be dimmed when/if needed.
❑ Sufficient light will be available so I can read my notes when the lights are dimmed.

Presentation Materials

❑ My notes are available, and I can read them while standing up.
❑ My visual aids are assembled in the order that I'll use them.
❑ A laser pointer or a wand will be available if needed.

Self

❑ I've practiced my delivery.
❑ I am comfortable with my presentation and visual aids.
❑ I am prepared to answer questions.
❑ I can dress appropriate for the situation.

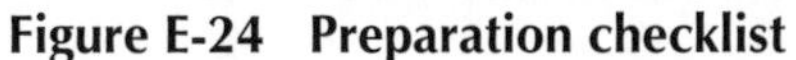
Figure E-24 Preparation checklist

Evaluating Access Presentations

Course: _______ **Speaker:** _______________________ **Date:** _______

Rate the presentaton by these criteria:
4=Outstanding 3=Good 2=Adequate 1=Needs Improvement
N/A=Not Applicable

Content

_____ The presentation contained a brief and effective introduction.
_____ Main ideas were easy to follow and understand.
_____ Explanation of database design was clear and logical.
_____ Explanation of using the form was easy to understand.
_____ Explanation of running the queries and their output was clear.
_____ Explanation of the report was clear, logical, and useful.
_____ Additional recommendations for database use were helpful.
_____ Visuals were appropriate for the audience and the task.
_____ Visuals were understandable, visible, and correct.
_____ The conclusion was satisfying and gave a sense of closure.

Delivery

_____ Was poised, confident, and in control of the audience
_____ Made eye contact
_____ Spoke clearly, distinctly, and naturally
_____ Avoided using slang and poor grammar
_____ Avoided distracting mannerisms
_____ Employed natural gestures
_____ Used visual aids with ease
_____ Was courteous and professional when answering questions
_____ Did not exceed time limit

Submitted by: ____________________________

Figure E-25 Form for evaluation of Access presentations

Evaluating Excel Presentations

Course: _______ **Speaker:** ________________________ **Date:** _______

Rate the presentaton by these criteria:
4=Outstanding 3=Good 2=Adequate 1=Needs Improvement
N/A=Not Applicable

Content

_____ The presentation contained a brief and effective introduction.
_____ The explanation of assumptions and goals was clear and logical.
_____ The explanation of software output was logically organized.
_____ The explanation of software output was thorough.
_____ Effective transitions linked main ideas.
_____ Solid facts supported final recommendations.
_____ Visuals were appropriate for the audience and the task.
_____ Visuals were understandable, visible, and correct.
_____ The conclusion was satisfying and gave a sense of closure.

Delivery

_____ Was poised, confident, and in control of the audience
_____ Made eye contact
_____ Spoke clearly, distinctly, and naturally
_____ Avoided using slang and poor grammar
_____ Avoided distracting mannerisms
_____ Employed natural gestures
_____ Used visual aids with ease
_____ Was courteous and professional when answering questions
_____ Did not exceed time limit

Submitted by: ______________________________

Figure E-26 Form for evaluation of Excel presentations

Index

C

D

E

J

K

L

M

N

R

S

T

U

V

W